Rebuilding Fisheries

THE WAY FORWARD

OECD

This work is published on the responsibility of the Secretary-General of the OECD. The opinions expressed and arguments employed herein do not necessarily reflect the official views of the Organisation or of the governments of its member countries.

This document and any map included herein are without prejudice to the status of or sovereignty over any territory, to the delimitation of international frontiers and boundaries and to the name of any territory, city or area.

Please cite this publication as:
OECD (2012), *Rebuilding Fisheries: The Way Forward*, OECD Publishing.
http://dx.doi.org/10.1787/9789264176935-en

ISBN 978-92-64-17692-8 (print)
ISBN 978-92-64-17693-5 (PDF)

Foreword

Many fisheries around the world are characterised by excessive fishing effort, low productivity and inadequate profitability. Considerable benefits can be made from rebuilding such fisheries. This publication analyses the issues and challenges governments face as they develop and implement plans to rebuild fisheries. The focus is on the economic and institutional issues and builds on evidence from OECD fisheries.

This publication is divided into five chapters. The first chapter presents principles and guidelines on the design and implementation of policies to rebuild fisheries. The second chapter examines the key factors that affect fisheries rebuilding plans and develops a framework for the analysis by identifying the tools, policies and pathways of rebuilding. Chapter 3 brings together the key results and insights from the case studies undertaken as part of the project.

Chapter 4 brings together information on rebuilding policies and measures undertaken by countries at the national and regional levels. This information is drawn from country surveys and also includes information for five Regional Fisheries Management Organisations. While this work was based on data available in 2010, the lessons learned and the on-going work of rebuilding fisheries that still is necessary remain valid today. The final chapter provides guidance on undertaking policy reforms needed for the rebuilding of fisheries.

Table of contents

Tables

Figures

Abbreviations

ACFA	Advisory Committee on Fisheries and Aquaculture
AFMA	Australian Fisheries Management Authority
ACFA	Advisory Committee on Fisheries and Aquaculture
CCSBT	Commission for the Conservation of he Southern Bluefin Tuna
CQs	Community based quotas
CFP	Common Fisheries Policy
IQs	Individual quotas
DFO	Fisheries and Oceans Canada
EPBC	The Environment Protection and Biodiversity Conservation Act
FMP	Fisheries Management Plan
FRMA	Fisheries Resources Management Act
FA act	Fisheries Administration Act
HSP	Harvesting Strategy Policy
ITQs	Individual transferable quotas
EFF	European Fisheries Fund
EEZ	Exclusive Economic Zone
LAPP	Limited Access Privileges Programmes
MEY	Maximum Economic Yield
MPA	Marine Protected Area
MSC	Marine Stewardship Council
MRAG	Marine Resources Assessment Group
MSE	Management Strategy Evaluation
MSY	Maximum Sustainable Yield
NAFO	Northwest Atlantic Fisheries Organization
RBM	Rights Based Management System
NGO	Non-Governmental Organisation
NOAA	National Oceanic and Atmospheric Administration
QMS	Quota Management System
RAC	Regional Advisory Council
TAC	Total Allowable Catch
RFMO	Regional Fisheries Management Organisation

UNCOVER	Understanding the Mechanisms for Stock Recovery
UNFSA	United Nations Fish Stocks Agreement
RBM	Rights Based Management
SARA	Species at Risk Act
STECF	Scientific, Technical and Economic Committee for Fisheries
TAC	Total Allowable Catch
TIS	Trade Information Scheme
TURFs	Territorial Use Rigths

Chapter 1.

Principles and guidelines on rebuilding fisheries

Rebuilding fisheries is an urgent task which is high on the international policy agenda. The OECD Committee for Fisheries decided to contribute to rebuilding efforts by providing an analysis of the main policy issues and challenges. The focus is on "rebuilding fisheries", which is a broader approach than "rebuilding fish stocks" and encompasses the social, economic and environmental dimensions of fisheries. The outcome of this project is a set of principles and guidelines that can assist policy makers in their efforts to make fisheries successful. These principles and guidelines aim to ensure that rebuilding plans are examples of good governance which implies inclusiveness, empowerment, transparency, and flexibility underpinned by predictable rules and processes.

The OECD Committee for Fisheries (COFI) decided in 2008 to contribute to efforts by member states to rebuild overfished and depleted fish stocks by investigating the economic aspects of fisheries rebuilding. This also contributed to the Organisation's on-going work on green growth and food security.

The conclusions of this work provide policy makers with a set of practical and evidence-based principles and guidelines to consider when designing and implementing rebuilding plans, whilst preserving the livelihoods that depend directly or indirectly on this activity. These principles and guidelines were adopted as a Council Recommendation by the OECD (OECD, 2012) in April 2012.

The present analysis focuses on a broad concept of rebuilding fisheries that goes beyond just the rebuilding of fish stocks. "Rebuilding fisheries" refers here to programmes (government-sponsored or otherwise) that seek to improve at once the stock status as well as securing both the integrity of ecosystems and livelihoods that depend on fisheries. An improved understanding of the economic, social and institutional issues that underlie successful rebuilding efforts will increase the likelihood that fisheries rebuilding programmes will meet their objectives.

This chapter outlines the motivation to rebuild fisheries; provides the general principles that underpin rebuilding, and outlines specific guidelines on the design, implementation and governance issues of rebuilding plans.

There is a need for action at all levels to ensure the long-term sustainable use and management of fisheries resources. Rebuilding fisheries is potentially both economically and socially beneficial as it:

- leads to a sustainable fishery where the harvesting and processing capacity is commensurate with the productivity of healthy fish stocks, thereby sustaining fishing communities, generating employment, and preventing a waste of human and physical capital;

- can increase food security and contribute to green growth; and

- has positive environmental effects, including the rebuilding of target fish stocks, supporting biodiversity, and strengthening the resilience of the ecosystem as a whole.

Principles

Fisheries should be managed in a sustainable and responsible way so as not to lead to a situation where rebuilding becomes necessary. Rebuilding plans should be based on social, biological and economic principles which should be incorporated throughout the design and implementation process in an integrated fashion, as opposed to sequentially or in isolation. Addressing risk and uncertainties should be explicitly incorporated into the rebuilding plan.

Efforts to rebuild fisheries should aim at restoring a sustainable fishery with a potential to generate profits and employment. Careful considerations of costs and benefits and their distribution is an important policy issue.

Efforts to rebuild fisheries should take into account relevant international fisheries instruments, as well as environmental and ecosystem considerations and the interactions between the fishing activity and other industries.

Rebuilding plans should be an integral part of a coherent broader fisheries management system. The management instruments employed should be consistent among themselves and consistent with instruments applied elsewhere in the management system.

Good governance, which implies inclusiveness, empowerment, transparency, flexibility and a predictable set of rules and processes for fisheries management, is a key element in ensuring success. Good governance acknowledges the tensions and balance between objectives of different stakeholders and contributes to resolving those tensions. Transparency helps to build trust and foster dialogue among stakeholders. The inclusion of a wide range of stakeholders (including different levels of government, environmental and scientific communities, industries and local communities) calls for a clear specification of each group's role in institutional structures and processes.

Guidelines

Rebuilding plans should be based on a comprehensive assessment of ecological, economic and social conditions, the interplay between fishing activity and the fish stock, and the existing management and governance regime while accounting for uncertainty. New and existing research, data and analysis can contribute to this assessment.

Rebuilding plans should have well-defined objectives, targets, harvest control rules and assessment indicators which are clearly articulated and measurable. The rebuilding plans should provide estimates of the time pattern of likely economic benefits and costs with respect to catches, capacity, profitability, distribution of added catch value, employment, over the time of the recovery period and these variables should be monitored during implementation. The original estimates and the results of the monitoring should be provided to stakeholders throughout the process in a clear and transparent manner.

Rebuilding plans should take account of the full costs and benefits of designing, implementing, and monitoring the programme, and their distribution

The design of rebuilding plans should take into account the characteristics of the fishery, such as fleet composition, the biological characteristics of the resource and whether the resources are managed at a local, national, regional or multilateral level.

In rebuilding plans, appropriate monitoring, control and surveillance instruments are necessary for successful implementation and should be designed and implemented for operational effectiveness, but should also address administrative simplicity and cost effectiveness.

Stakeholders have an important role to play in many stages of the rebuilding process to ensure a common understanding of the state of the fishery. Such engagement will help in the development of clear, transparent policies that provide managers and stakeholders with a degree of predictability with respect to process and expected changes in policy variables, and may therefore help build support for rebuilding.

Fisheries rebuilding plans should be communicated to the general public and results of their implementation reported in a timely fashion.

Rebuilding often implies incurring short-term costs in the interest of generating long-term benefits; weighing these costs and benefits is an important undertaking. The distribution of cost and benefits among stakeholders is a key policy consideration and will significantly influence stakeholders' support for a plan. Rebuilding plans should therefore: clearly articulate expected costs, benefits, and their distribution in the short and long term; seek to ensure that those stakeholders who bear the costs of rebuilding will receive some of the benefits; and should be designed to allow stakeholders to better recognise and value the expected long-term benefits of rebuilding efforts.

Rebuilding plans should account for the interaction between central and local authorities as well as a broad range of stakeholder groups. Decisions taken at the local level influence decisions taken at higher levels, and *vice versa*. This interaction should be addressed in the rebuilding plan and in the governance system more broadly.

The implications of risk and uncertainties, and means to address them and where possible reduce them, should be explicitly incorporated into the rebuilding plans. Rebuilding plans should be robust and adaptive to variability and unexpected changes in the environment, industry or the economy. The design of rebuilding plans should include mechanisms to monitor progress and anticipate actions to be taken if rebuilding is not advancing. It is important to have a mechanism to assess and communicate to the stakeholders and policy makers the biological and economic risks associated with various components of the rebuilding plan. Mechanisms that take uncertainty and risk explicitly into account and reduce possible negative effects should be used.

Rebuilding fisheries usually requires the concurrent use of multiple management measures. Measures may include input/output controls as well as various technical measures. Generally, output controls are effective in restraining catches but can be costly to enforce and monitor. Input controls are often less effective in restraining catches but may be cheaper and easier to implement.

Rebuilding requires a modification of fishing mortality to increase stock sizes and improve stock structures, and the management instruments in use should be effective in this regard.

When a rebuilding plan concerns a species found in a multispecies, multi-gear fishery, specific management measures should be applied due to the interactions between the gears and fisheries, and the possible effects that this particular rebuilding initiative may have on other species and fisheries should be addressed.

Rebuilding plans should take account of by-catch and discards, and include measures to reduce these where possible.

Habitat conservation and enhancement can be an important part of rebuilding plans.

The pace of rebuilding is an important aspect of a rebuilding plan. A moratorium or a sharp reduction in effort or catch can result in idled human and physical capital with accompanying waste and lost know-how and markets. Higher net present value of fishery output will normally be achieved by reduced but positive harvest levels, although this may require a longer time period to achieve the targets. In many cases a gradual or incremental implementation of the rebuilding plan can be useful as it may help to increase social acceptability, prevent abrupt economic and social harm, and ease the financial and political pressures on governments. However, this gradual approach must be balanced against the possibility of significant and potentially irreversible damage to the fish stock and/or the ecosystem if harvest continues.

Retraining programmes, well-designed decommissioning schemes[1] and other flanking measures may help stakeholders to adapt to the changes in the fishery. Such measures may also engender stakeholder support for the rebuilding plan.

Harvest control rules or similar measures, where applicable, are central to rebuilding fisheries. They specify predetermined management actions, especially those related to allowable harvest levels, according to the difference between the current stock size and structure and target stock objectives. The use of such rules also allows for discussing and agreeing on specific trajectories, taking into account possible social and economic impacts and uncertainties.

Experience shows that there are various types of individual and collective rights-based management instruments that may be useful to consider under different conditions by creating incentives for industry self-adaptation. Well designed rights-based management systems may be effective if the objective is to reduce fishing effort, while at the same time securing profits for fishers in the longer term. Challenges associated with rights-based management can be addressed through specific safeguarding measures.

An integral part of a rebuilding plan is to decide on how the fishery shall be managed after the rebuilding period. Such a post rebuilding plan should ideally secure a sustainable fishery and prevent back-sliding.

Note

1. See, *inter alia*, the *Recommendation of the Council on the principles and guidelines for the design and implementation of plans for rebuilding fisheries* (OECD, 2012).

Chapter 2.

Why rebuild fisheries and how

From a biological, environmental and socio-economic perspective, many of the world's fisheries are in poor condition. Rebuilding and managing fisheries in a biologically and environmentally sustainable way can bring considerable social and economic benefits. Policy makers are therefore pressed to rebuild fisheries. A rebuilding plan begins with evaluating the state of the fishery including the environmental and socio-economic situation. The next steps include setting feasible rebuilding goals, deciding on mechanisms to achieve them, monitoring progress, and ensuring the long-term sustainability of the fishery once rebuilt. Special attention must be given to risk and uncertainty in rebuilding plans. In this regard, applying risk evaluation and communication of risks to stakeholders is important. Policy makers have many management tools to rebuild fisheries. The tools used will depend on the specific characteristic of each fishery; in all cases, however, a mix of tools is needed to successfully rebuild a fishery.

The so-called fishery problem can be stated thus: There are too many fishers, each operating without sufficient constraints on their individual behaviour and chasing too few fish. Although this may be an oversimplification with regards to many fisheries, it is nevertheless true for too many fisheries. It has been estimated that the rent dissipation in world fisheries amounts to USD 50 billion per year (World Bank, 2008). A part of this rent dissipation derives from the fact that many fish stocks are below their optimal size or that excessive resources, such as labour, vessels or capital, are used to harvest stocks. Although much work is needed to prevent more fisheries from being overfished, there is also a need to look closely into how fisheries can be rebuilt.

Considerable benefits can be made from rebuilding fisheries. These include monetary gains (Costello *et al.,* 2012, Salz *et al.,* 2010), but also contributions to improved social and environmental outcomes by increasing employment, securing livelihoods, maintaining biodiversity, and providing stable and safe food supplies. From an ecosystem viewpoint, rebuilt fisheries contribute to the sustainability and stability of ecosystems making them less vulnerable to variations in the environment and to outside shocks.

In many cases, the rebuilding objectives are defined solely in biological terms such as maximum sustainable yield (MSY), and the design of the paths and instruments to achieve these objectives are often driven by biological considerations. This overlooks the reality that rebuilding programmes are nested within a broader economic, social and political setting. In this context, it is worth noting that a useful distinction can be drawn between *rebuilding fish stocks* and *rebuilding fisheries*; the former is focused more or less exclusively on the species and its habitat, while the latter extends to the fishing industry and associated communities so as to capture the human dimension as well. *Rebuilding fisheries* acknowledges the social, economic and governance components of rebuilding.

This publication focuses on cases where stocks are significantly depleted and there is a desire to rebuild them. It should be noted that according to the terminology used here rebuilding assumes that socio-economic objectives are taken into consideration in the design and implementation of the rebuilding plans.

While biologically defined objectives and commitments such as that made at the World Summit of Sustainable Development (WSSD) represent a crucial pledge to rebuild fish stocks, the question of how economic analysis and instruments can be applied towards rebuilding fisheries, including the development of rebuilding objectives and timeframes, remains a challenge. A holistic approach emphasises that economics has a crucial role to play in influencing the choice of rebuilding objectives, the rebuilding path, the technical and policy instruments used to achieve objectives, and the enforcement mechanisms.

The task of rebuilding fisheries is a challenging policy problem. Rebuilding fisheries is not solely a technical issue, such as lowering fishing mortality to a level that ensures the recovery of fish stocks. Rebuilding is usually a complicated process requiring co-operation between the different stakeholders; that is, policy makers, scientists, government officials, the industry, and communities. Indeed, fisheries management and rebuilding plans are more about managing people than about managing fish (Hilborn, 2007; Davis, 2010). In this regard, fisheries managers and policy makers would do well to focus on the fact that fishers often have an incentive to overfish (OECD, 1997).

As mentioned in Chapter 1, the focus here is on the economic and institutional aspects of the rebuilding process and the analysis builds on previous OECD work on market-based instruments and the political economy of reform, in addition to a number of case studies. The case study material covers rebuilding efforts for fisheries in OECD countries and selected developing countries, as well as from fisheries managed by RFMOs. These case studies cover a range of different economic, environmental and policy contexts. The *OECD Workshop on the Economics of Rebuilding Fisheries*, held in Newport, Rhode Island in 2009, also provided valuable input.[1]

It should be noted that the background for fisheries management and rebuilding across countries is not uniform. The different approaches used are highlighted in the *Inventory of National and Regional Approaches to Fisheries Rebuilding Programmes* compiled for the purposes of this project. Institutional arrangements, legislative requirements, decision-making structures and stakeholder involvement vary considerably. They range from specific laws that mandate exact timeframes and procedures to more flexible approaches that are based on policies and guidelines. By extension, this also means there is no one-size-fits-all solution; approaches need to be tailored to specific circumstances.

The institutional and legislative structures of a country or region have a role to play. In some countries, stock rebuilding is mandated as part of the national fisheries management legislation; this has led to a prescriptive approach with tight timelines and limited flexibility. In other countries, rebuilding plans are more *ad hoc* dealing with specific fisheries using tailor-made instruments and regulations. Analyses of the case studies indicate there is a higher likelihood of success in jurisdictions that have specific and strict legislation that govern stock building (Caddy and Agnew, 2004; Wakeford *et al.*, 2007).

The framework for rebuilding

This chapter highlights the main issues regarding the need to rebuild fisheries and examine some of the possible benefits and challenges encountered to date. It begins by examining the motivations for rebuilding and describes the present situation with regards to the need for rebuilding. This is followed by an overview of the different possible trajectories for rebuilding fisheries and some of the major issues related to how these trajectories should be compared between fisheries and over time. Several challenges related to uncertainty and risks in rebuilding are also examined.

The motivations to rebuild

Efforts to halt overfishing and rebuild depleted fisheries have received increasing attention in the last few years, resulting in pressure on governments to take stronger action at the international, regional and national levels. This attention has revealed the poor state of many fisheries (Box 2.1), and an increasing and vocal number of stakeholders are seeking to ensure better fisheries management and governance.

The trend is clear. A number of fish stocks around the world have been depleted with implications for biological sustainability as stocks and/or the ecosystems are at risk, and economic prosperity, as fishing activity directed at depleted stocks is inefficient and involves the waste of inputs. The collapse of several high profile fish stocks, such as the Northwest Atlantic cod, and their failure to recover despite a reduction or moratorium on fishing effort has raised concerns over the success of recovery plans for overfished stocks (Caddy and Agnew, 2004; Rosenberg and Mogensen, 2007; Wakeford *et al.*, 2007). The economics of overexploitation has also been widely studied (Grafton *et al.*, 2007).

The situation is not all negative, however. Several countries have taken steps to rebuild fisheries with different levels of success. The Canadian halibut fishery and the sablefish fisheries, for example, were successfully re-established. Initial attempts to use input controls (mostly days at sea regulation) to diminish fishing mortality were unsuccessful as the fishing seasons imposed were short and the inputs wasted when fishers raced to gather as much fish as possible. Short fishing seasons also led to inefficiencies in the processing and marketing sectors. The introduction of individual transferable quotas in these fisheries, however, resulted in a longer season and rent generation (Munro, 2010).

Japan and Korea used different measures, relying on active stakeholder participation though community management structures and self-imposed management measures, such as time and area closures (Uchida, 2009; Lee, 2009, Uchida *et al*., 2010). Although the economic results are mixed and it is too early to know the outcome, there are positive signs with regards to stocks. Iceland has managed to retain a profitable and efficient cod fishery in spite of drastically reducing harvest rates. The same can be said of the Danish Baltic Sea cod fishery. As a result of policy changes mandated by the Revised Magnuson Stevens Act, including the application of strict and conservative harvest control rules to set annual catch limits, the United States is making good progress to ending overfishing in domestic waters. Between 2000 and 2010, 84 stocks were on the overfished list; in that same period, 36 stocks came off the list. At the same time, there were 76 stocks subject to overfishing; 36 stocks have come off that list.

Apart from the apparent poor state of many stocks, there are other arguments in support of fisheries rebuilding which can be classified under three different headings; *environmental, economic* and *social*.

From an *environmental and ecosystem* perspective, rebuilding a fishery is necessary to secure biodiversity and the resilience of the ecosystem. High fishing mortality and excessive fishing effort can to the fishing activity no longer being viable. This may also have a negative effect on the habitat and harmful effects for other living organisms in the ecosystem.

From an *economic* viewpoint where excess harvesting has lead to low harvest levels, even though the fishery could still be biologically sustainable there is a waste of economic potential from that fishery. Such fisheries are often characterised by low profits for harvesting firms and low incomes for fishers.

Dwindling and/or fluctuating stocks and catches create problems for processing firms, markets and the value chain more generally as supply is unstable. Fluctuations in supply and quality make it difficult for retailers and consumers to evaluate the product offered to them. It can also increase costs in the value chain because of the challenging and complex logistics. Therefore, fish from capture fisheries are often at a disadvantage when competing with other food products and aquaculture in particular. Rebuilding plans may also aggravate a fragile market situation during the transitional period as fluctuating or dwindling supplies may result in a loss of market. The Japanese, Korean and Estonian case studies illustrate this point. Such effects may discourage stakeholders to engage in rebuilding as there may be uncertainty about market access once the fishery is rebuilt. The case of the Norwegian spring spawning herring stock, where a minor fishery was kept open during rebuilding, is informative in this regard. Although the stock may have recovered more quickly with a moratorium in place, keeping the fishery open with a small allowable catch seems to have paid off. A moratorium would most likely have resulted in loss of market access (Sandberg, 2009).

Box 2.1. The state of the world fisheries

It is technically difficult to assess the state of the world's fisheries. The FAO uses the following classifications for stock status.

- *Underexploited, undeveloped or new fishery.* Believed to have a significant potential for expansion in total production.

- *Moderately exploited, exploited with a low level of fishing effort.* Believed to have some limited potential for expansion in total production.

- *Fully exploited.* The fishery is operating at or close to an optimal yield level, with no expected room for further expansion.

- *Overexploited.* The fishery is being exploited at above a level which is believed to be sustainable in the long term, with no potential room for further expansion and a higher risk of stock depletion/collapse.

- *Depleted.* Catches are well below historical levels, irrespective of the amount of fishing effort exerted.

- *Recovering.* Catches are again increasing after having been depleted or having suffered a collapse from a previous high.

The figure below shows the trends in the world stocks according to this classification. It suggests that the situation is serious especially with regards to the number of species that are overexploited, depleted or recovering. There are, however, positive signs. Around 1995, the proportion of stocks harvested around MSY increased. The proportion of stocks recovering (not shown explicitly on the graph) has increased, mostly in recent years. However, the number of stocks offering potential for expansion is decreasing, clearly illustrating there is little scope for increasing world catches. There are, however, certain caveats regarding these data.

First, the assessment is based on data for specific species using FAO's statistical areas which in most cases do not coincide with areas of national or international jurisdiction. FAO statistical areas are mostly based on geographical rectangles, whereas areas of national jurisdiction are defined by a virtual extension of the coastline. Thus, within an FAO statistical area, there can be a number of EEZ's or an FAO statistical area may cover areas which are subject to regulations by more than one RFMO.

Secondly, each assessment may include several species groups, and each of these may include numerous stocks or sub-stocks. Thus, trends of individual stocks or sub-stocks may be masked by the overall pattern of the larger group within which it is contained.

Finally, the classification of stock status is primarily based on stock abundance, but as information on abundance is not available for all species other indicators may be used. The methodology used may therefore differ from one species to another.

Figure 2.1. Global trends in the state of exploitation of the world's fish stocks, 1974-2004

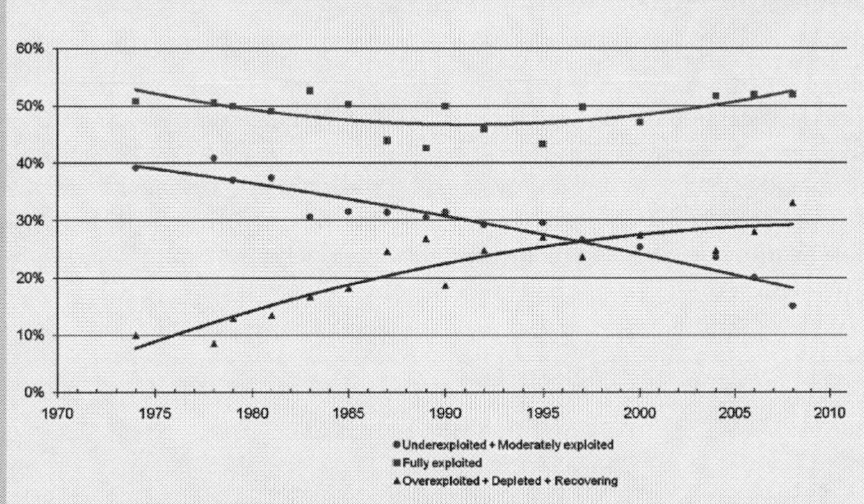

Source: Based on FAO data.

Negative *social* impacts of excess harvesting and effort in fisheries are numerous. Employment fluctuates, working conditions are difficult, and worker safety is often compromised. Larger and more stable catches benefit fishing communities, especially where alternative employment opportunities are rare.

Declining stocks may have other social implications such as loss of a fishing culture, know-how and expertise which can be difficult to maintain if the fishing activity is suspended for a long period. At the regional level, the socio-economic benefits from rebuilding fisheries include protecting a way of life and maintaining employment in coastal communities. In this context, stable access to well-managed resource decreases economic uncertainty and secures harvesters' access to capital. This allows the industry to remain competitive while leading to the production of higher quality, higher value fish products.

When rebuilding plans imply lowering fishing mortality through decreased fishing effort, the immediate effects are negative for fishers and those working in the value chain, e.g. less income and employment. How to deal with these effects is an essential part of any rebuilding plan. The engagement of stakeholders and transparency on anticipated trade-offs between short-term losses and a medium- or long-term gain is necessary. The likelihood of a rebuilding plan being successful depends largely on the willingness of stakeholders to support and participate in the plan.

Although economic, social and environmental considerations of rebuilding fisheries are important it is worth noting that many of the decisions are taken by policy makers. In most countries, fisheries are managed by laws, rules and regulations which have been developed and implemented by the political establishment. Rebuilding fisheries, as part of fisheries management, is therefore in most cases a policy problem that addresses economic, social and environmental concerns.

In addition, numerous countries are bound by international law (United Nations Convention on the Law of the Sea), international treaties, and national legislation to manage their fisheries in a sustainable and responsible way. In 2002, governments agreed at the World Summit on Sustainable Development (WSSD) to rebuild fish stocks by committing to "Maintain or restore stocks to levels that can produce the maximum sustainable yield with the aim of achieving these goals for depleted stocks on an urgent basis and where possible not later than 2015" (UN, 2002).[2]

Given the costs and benefits described above, policy makers are faced with a variety of options. In many cases the benefits of rebuilding are greater than the costs, but their distribution in time and among stakeholders varies. Hence, managing the transition from the depleted state to a rebuilt one, as will be discussed later, is a key to success. However, weighing into the decisions by policy makers and fisheries managers will be lobbying efforts by stakeholders. For most of the analysis that follows, it is assumed that the decision to rebuild has been taken, and that the focus is on the design and implementation of the plan itself.

Steps to rebuilding

A rebuilding plan is a process. Irrespective of the specific design of a rebuilding plan the process can be divided into several steps as shown in Figure 2.2.

Figure 2.2. Steps of a rebuilding fisheries plan

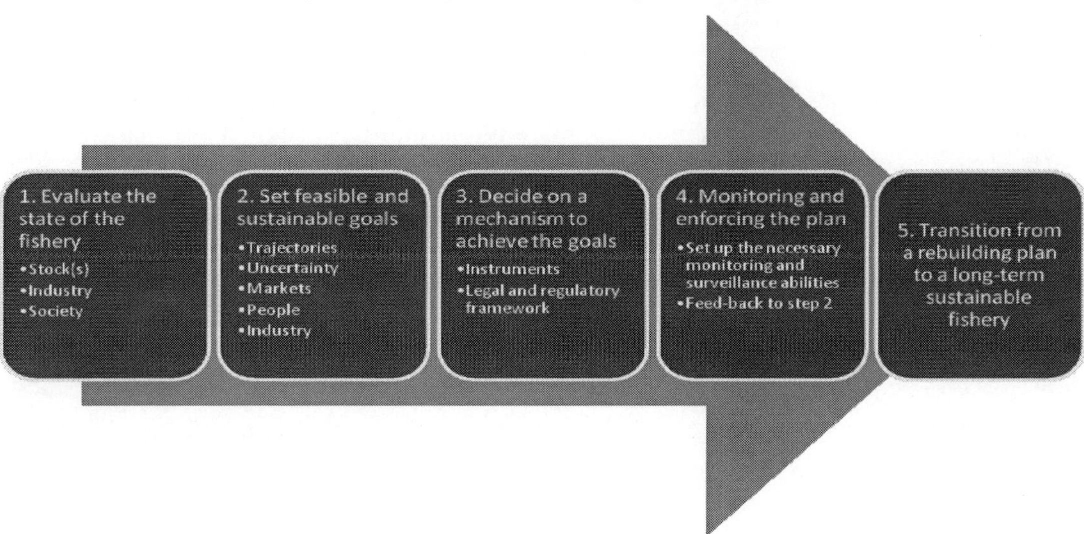

The *first step* is to evaluate biological and environmental factors, as well as the state of the fishing industry and the fishing community. In order to simplify the analysis, it is taken as given that policy makers have decided to rebuild a specific fishery. Once this decision is made, it is necessary to obtain relevant data and perhaps to acquire new information regarding the fishery. Balancing costs and benefits of acquiring new information may be an important consideration.

Step two is setting feasible and sustainable objectives. Questions regarding how quickly the fishery should be rebuilt, where uncertainties lie, and how markets, people and industry will be affected must be addressed. Identifying relevant stakeholders, mapping out their distinctive role in the rebuilding plan, and addressing the distribution of costs and benefits are equally important. Great care must be given to aligning the objectives of the rebuilding plan with the interests of stakeholders.

The *third step* is to decide on the mechanisms to achieve the stated rebuilding objective. This includes the choice of management instruments and the regulatory framework for the rebuilding plan. Such choices may include a mix of input and output controls, seasonal and area closures, and the allocation of rights either to communities, individuals or linked to specific areas. The necessary rules and regulations must be in place, as is a monitoring and surveillance system that ensures the rules are followed. The extent of monitoring activity may differ from one case to another. In some cases, it may be possible to leave it to the stakeholders to undertake or strengthen some of this activity, as is often the case in rights-based management systems.

Step four is to determine a mechanism to monitor the performance of the plan. This entails collecting and disseminating information on key indicators on the successes or failures of the plan and how it can be strengthened. It is also necessary to have a mechanism that allows the rebuilding plan to adapt to changes. For example, if the biological or socio-economic circumstances change, it is necessary to be able to reevaluate steps 1 and 2. Transparency and measurable indicators play a key role in this regard.

The *fifth step* is to set up a post-rebuilding fisheries management regime. Given that the rebuilding effort imposes costs and hardships on the industry, the active participation of all stakeholders is crucial and hence stakeholders must secure some of the benefits that rebuilding creates.[3] Step five should also ensure that the fishery does not revert to its former state.

The five steps shown in Figure 2.2 are interdependent and have feedbacks and loops between them. To give an example, the monitoring and enforcement plan chosen may affect the objectives and the mechanisms used to attain them. Given such interdependencies, it is necessary to integrate the different steps when designing the overall rebuilding plan.

Sustainable rebuilding

Although this study looks at the economics of rebuilding from a wide perspective, it focuses on fisheries facing challenges due, in part, to the fact that the fish stock or stocks are too small.

Without addressing the issue of whether the maximum sustainable yield is a sensible or even an attainable objective, there is a long-run average relationship between stock size and sustainable harvest levels; up to a certain point increases in stock size lead to an increase in sustainable yield (Figure 2.3). This figure shows the sustainable yield curve, i.e. each point on the curve shows the sustainable (equilibrium) yield from the specified stock size.

Figure 2.3. The relationship between biomass and sustainable harvest

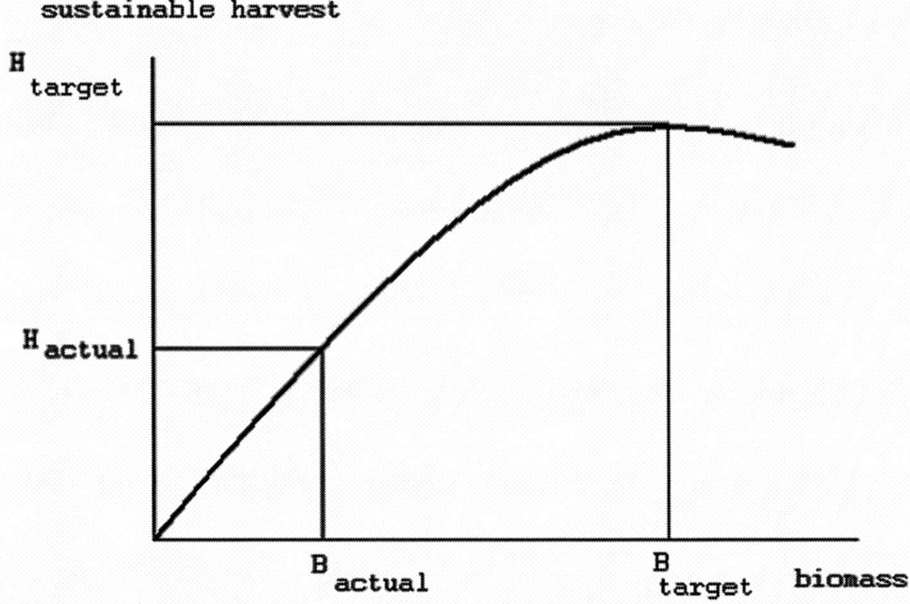

For ease of exposition, Figure 2.3 has been drawn so that current biomass is lower than what policy makers would consider to be optimal. The result is a harvest lower than that obtained if the biomass was greater, which is a common feature for fisheries in need of rebuilding. The policy objective is to lower the harvest rate or effort to allow the stock to grow until it reaches the target biomass level (this dynamic adjustment is not shown in Figure 2.3). The point is that given the general assumptions about the relationship between biomass and sustainable harvests it is possible to reach a sustainable state of nature where the stock is larger and the harvest greater than it is at present.

Ecosystem considerations

The analysis above is based on the simplified assumption that fishing mortality is the only factor which affects stock size. The ecosystem within which fisheries takes place is more complex and there are other factors that come into play. This includes temperature, currents, salinity, and habitat. For example, climatic variation may greatly affect the size, growth and behaviour of fish stocks. In addition, the time scale of such changes affects rebuilding plans. If climatic changes and subsequent changes in the aquatic environment are important and rapid enough they may undermine efforts to rebuild fish stocks.

A related issue is changes in habitat, caused for example by human activities. The fishing activity itself affects the habitat and will differ from one fishery to another, due to different circumstances, including the level of fishing effort and intensity, gear types and fishing methods. Furthermore, fisheries operate in a space which is used for many other activities other than fishing. Transport, tourism, oil extraction and gravel extraction are examples of economic activities which may directly and indirectly affect the ecosystem, and impact the fish stocks. It is worth noting that some habitats are particularly vulnerable to exogenous impacts.

Most fisheries are multi-species fisheries as most fishing techniques have limited selectivity. For this reason, by-catches are common and can complicate rebuilding efforts. Fishers may be less inclined to take part in a rebuilding plan if they may not fish more of the successfully rebuilt species because of excessive by-catch of some other species. Indeed, given the complex set of interactions between species, the depletion of one stock may have consequences for other aquatic organisms and, more generally, the ecosystem. For example, the abundance of northern shrimp and crab stocks in the North East Atlantic may be a result of the overfishing of their predator, Atlantic cod (in addition to potential changes in oceanic conditions). The cod-capelin-shrimp fisheries in Iceland are another case in point; it has proven difficult to manage those fisheries in a holistic fashion although considerable effort has been given to mapping the species interactions (Jakobsson and Stefansson, 1998).

Furthermore, predator-prey relationships greatly complicate rebuilding efforts and are often complicated, non-linear and difficult to predict. They are, however, important in many fisheries. For example, large sharks (which are slow maturing and long-living species) play a key role as predator in the ecosystem. The collapse of many shark populations off the north eastern coast of the United States has led to an increase in the number of skates and rays, species that were typical prey of sharks; this has had significant consequences on fishing communities as the increased number of skates and rays have affected commercial fisheries for bay scallops and other shellfish (Myers *et al.*, 2007).

Protecting biodiversity is an important ecological benefit of stock rebuilding. Diverse ecosystems are considered to be more resilient and have a greater ability to withstand changes in the environment, such as ecosystem shifts and climate change.[4] Some progress has been achieved going from single species to multi-species management but difficulties remain, especially due to the biological complexities and uncertainties inherent in such fisheries. Canada's Pacific Groundfish Integration Programme provides important insights into how quotas in non-targeted species can help rebuild fisheries which are both multi-gear and multi-species.

Despite these challenges, a more holistic ecosystem approach is necessary as to focus on single species management can have unforeseen and possibly negative consequences. While it may be difficult to employ an ecosystem approach using current quantitative models, great strides can be made by using this approach in qualitative models, e.g. in planning. This approach greatly increases awareness of the interactive forces at work in the ecosystem and may increase the robustness of the rebuilding plans. Limiting efforts to single-species considerations in the design and implementation of rebuilding plans opens the risk of unforeseen complications, and failure may be more likely than when taking an ecosystem approach.

These and other issues related to the challenging complexities and interdependencies in ecosystems highlight the need to advance the ecosystem agenda. Many countries have taken steps towards an ecosystem approach to fisheries management, but little has been done to analyse how such an approach can help to rebuild fisheries.[5]

The aim of rebuilding plans

As previously stated, this study is concerned with the problem of rebuilding fisheries rather than simply rebuilding fish stocks. There is more than a subtle difference between the two approaches. If the aim is simply to rebuild fish stocks from a purely biological perspective, the most effective way is usually to stop fishing, possibly combined with other biological initiatives such as stock or habitat enhancement. However, if the aim is to rebuild a fishery while dealing with the economic, social and environmental consequences then economic considerations must be taken into account. Otherwise it is not possible to provide benchmarks for time-paths, adjustment measures, and other building blocks of the rebuilding plan. Biological considerations are inherent in all socio-economic based rebuilding plans as it is not possible to design such plans without taking the biological characteristics of the fishery into consideration.

The issue of definitions

An important issue concerns the definitions of key concepts such as overfishing and overfished, as well as what is meant by "rebuilding" and "success". Several countries and international organisations have developed broad definitions of overfishing that are generally linked to the maximum sustainable yield (MSY) concept, while others have not (Box 2.2). In the case of the New Zealand Ministry of Fisheries, a specific distinction is drawn between overfished and depleted stocks; this is an explicit recognition that in some cases ecosystem shifts or climate change may be a significant factor that affects the rebuilding of some stocks. In the United States, the Magnusson Stevenson Act states that overfishing is a rate or level of fishing mortality that jeopardises a fishery's capacity to produce MSY on an ongoing basis.

Box 2.2. National definitions of key terms

Several countries have defined key terms associated with fisheries rebuilding, either through legislation or through policies and guidelines. Below is an overview of several key terms.

New Zealand

Overfished: Stocks that are below a biomass limit, such as the soft limit, are frequently referred to as "overfished." However, the term "depleted" should generally be used in preference to "overfished" because stocks can become depleted through a combination of overfishing and environmental factors, and it is usually impossible to separate the two.

Overfishing is deemed to be occurring if FMSY (or relevant proxies) is exceeded on average. Rebuilding plan: A series of catch or fishing mortality levels designed to rebuild a depleted stock (i.e. a stock that has fallen below the soft limit) back to the target.

United States

Overfished: An overfished stock or stock complex "whose size is sufficiently small that a change in management practices is required to achieve an appropriate level and rate of rebuilding." A stock or stock complex is considered overfished when its biomass population size falls has declined below the level that jeopardises the capacity of the stock or stock complex to produce MSY on a continuing basis.

Overfishing: According to the National Standard Guidelines, "overfishing occurs whenever a stock or stock complex is subjected to a rate or level of fishing mortality that jeopardises the capacity of a stock or stock complex to produce maximum sustainable yield (MSY) on a continuing basis."

Rebuilding Plan: A management programme that increases a fish stock's biomass to an amount that can produce MSY on a continuing basis. Rebuilding plans should have defined starting and end dates, and specify a biomass target (Bmsy) it needs to reach by the end of the rebuilding period that can produce MSY. It should also specify interim target biomass amounts that should be reached during the rebuilding period so that managers can assess if adequate rebuilding progress is being made.

Australia

Overfished refers to the biomass of a fish stock. There are too few fish left; more technically, the stock has a biomass below the limit reference point. The Commonwealth Fisheries Harvest Strategy Policy (the policy) requires that fish stocks remain above a biomass level at which the risk to the stock is regarded as too high (BLIM or a proxy). Two common proxies for that limit are 0.5 BMSY (half the biomass required for maximum sustainable yield) and B20 (20% of the unfished biomass).

Overfishing refers to the amount of fishing. The stock is undergoing too much fishing; that is, the amount of fishing exceeds the limit reference point. The policy indicates that any directed fishing on an overfished stock amounts to overfishing. Fishing mortality (F) exceeds the limit reference point (FLIM). When stock levels are at or above BMSY, FMSY will be the default level for FLIM. Fishing mortality in excess of FLIM will not be defined as overfishing if a formal 'fish down' or similar strategy is in place for a stock and the stock remains above the target level (BTARG). When the stock is less than BMSY but greater than BLIM, FLIM will decrease in proportion to the level of biomass relative to BMSY. At these stock levels, fishing mortality in excess of the target reference point (FTARG) but less than FLIM may also be defined as overfishing, depending on the harvest strategy in place and/or recent trends in biomass levels. Any fishing mortality will be defined as overfishing if the stock level is below BLIM.

European Union

Overfishing is defined as "any fishery where the total fishing effort is greater than is required to meet or match a specific management objective, e.g. maximum sustainable yield (MSY)." A recovery plan is defined as "a set of measures aimed at rebuilding depleted stocks. Covering a period of several years, the plan is generally implemented in phases that can begin with emergency measures and the establishment of technical measures, as in the case of the recovery plans for cod and hake. All this is matched with monitoring and control and possibly even financial aid for the stakeholders concerned, to ensure that fishing pressure on the depleted stock is reduced."

Source: Country submissions from the European Union, New Zealand, United States.

There are two broad approaches to rebuilding fisheries. The first is the explicit use of targeted rebuilding programmes, focusing on a particular fishery or fisheries and employing specific measures to reduce fishing effort. Such programmes are generally time-limited, have legislative backing, and employ precautionary reference points and harvest control rules within the rebuilding framework (Box 2.2). The second approach includes rebuilding as one of several objectives within regular fisheries management plans. Such plans may include precautionary reference points and harvest control rules as part of the management approach, but these may not necessarily be mandated by legislation. This approach to rebuilding is more open-ended compared to targeted rebuilding programmes, and there may be a greater emphasis on the on-going management of the fishery both during and following the rebuilding portion of the programme.

For the purposes of this study, the term "rebuilding programmes" is used in a generic sense to encompass both targeted rebuilding programmes and rebuilding objectives included within broader fisheries management plans.

Wakeford *et al.* (2007) note there is an important technical and legal distinction to be made between *recovery* and *rebuilding* plans in some jurisdictions. Recovery plans are generally focused on the recovery of critically endangered species from the risk of extinction and does not necessarily imply rebuilding to commercially sustainable levels. Legislation to address this threat is generally found in the endangered species legislation of countries, rather than in the fisheries management legislation. Rebuilding plans, however, are generally associated with the rebuilding of the fishery to more productive levels of exploitation. In some countries, the endangered species legislation may serve as a catalyst for rebuilding fisheries and in these cases a *recovery* plan would be within the scope of this project. This study will focus on rebuilding plans regardless of their legislative basis.

Another definitional issue is what is meant by "success" in rebuilding programmes. In the long term, a rebuilding programme can be defined as a success when the target stock biomass is achieved and the conditions are established for the fishery to be economically and environmentally sustainable. However, such an objective can take a considerable amount of time, and governments may wish to have short term or intermediate objectives against which to define and measure success (or progress towards success over the longer term). In the short term, a rebuilding programme can be said to be successful if the actual fishing mortality rate is less than or equal to the target and the stock biomass is increasing. This places greater policy emphasis on the issue of biological overfishing in the short term and provides measurable intermediate objectives towards the longer term objective of a rebuilt fishery.

Within the scope of this study, success in rebuilding is measured against the objective of achieving an environmentally sustainable and economically viable fishery that maximises societal benefits (broadly defined) subject to resource constraints.

Choice of rebuilding targets

It is clear that sustainable socio-economic benefits from a fishery cannot be reaped unless the fish stock itself is at a biological sustainable level. As the focus is on fisheries that are below what is thought to be the desirable stock size, it is necessary to set clear stock targets given the biological characteristics of the species, and the economic and social characteristics of the fishery.

As the use of biological reference points is standard practice in most fisheries management and rebuilding plans, it is worthwhile to give an overview of the most common ones.

Biological aspects

Precautionary reference points are used in fish stock assessments and fisheries management to assess the long term sustainable levels of fishing mortality and stock biomass. Such reference points or similar tools are necessary inputs into the design of rebuilding strategies. The reference point approach is usually used to determine threshold or targets in the fisheries rebuilding plans. The following four reference points are normally used:

B_{lim}: (Danger level) scientists have proposed this as the limit below which there is a serious risk of stock collapse.

B_{pr}: (precautionary level) is set at a higher level which gives reasonable certainty that in spite of year-to-year fluctuations the stock will stay above B_{lim}.

F_{lim}: the level of fishing mortality at which there is an unacceptably high risk that the stock will collapse.

F_{pr}: a lower level of fishing mortality which offers a high probability that the stock biomass will stay above B_{lim}.

The starting point of the rebuilding plan is that the current fishery policy has lead to a degradation of the stock (with a high risk of falling below a threshold level (B_{lim}), where the reproduction level is impaired). It should be noted that there are often uncertainties about the actual size of the stock biomass; the level of B_{lim} and the reasons why the stock is at a low biomass, e.g. fishing mortality, natural variations, changes in the climatic trend, or some combination of these.

Estimates of biomass and targets are usually based on the history of catches, but these catches on their own are not a good estimator of the carrying capacity of stocks. In the literature on rebuilding, reference is often made to historically big catches which can become a sort of a reference point for fisheries managers. However, such historical peaks are usually unsustainable and often occur during periods of overfishing, and hence should not be used as targets.

There are difficulties in calculating biological reference points due to various uncertainties in data and model specifications. This does not mean that such difficulties should hinder fisheries managers in obtaining the best available estimates, but they should consider the message given by two prominent specialists in fisheries management: "If we have learned anything from the historical performance of fisheries management it is that it is more important that the basis for fishery management action to be clear and indisputable than that it should claim to be precise and accurate" (Caddy and Mahon, 1995).

Economic aspects

The fundamental economic problem with fisheries is depicted in Figure 2.4. It shows a theoretical revenue curve and a cost curve of general shapes. The cost curve shows economic cost, i.e. the opportunity cost associated with the fishing activity. The revenue curve is a function of the sustainable yield curve shown on Figure 2.3 and depicts the sustainable revenue for each effort level up to e_{oa}. Both the revenue and the cost curves reflect simple but common biological characteristics. This is a very general and theoretical description of many real-world fisheries and is presented to give a rough overview of the fundamental economic problem of fisheries management and rebuilding fisheries. Different assumptions concerning the biological and economic characteristics of specific fisheries may change the shape and position of the curves, but the model is nevertheless robust for most fisheries (Larkin *et al*, 2011).

Figure 2.4. The socio-economic problem of open access fisheries

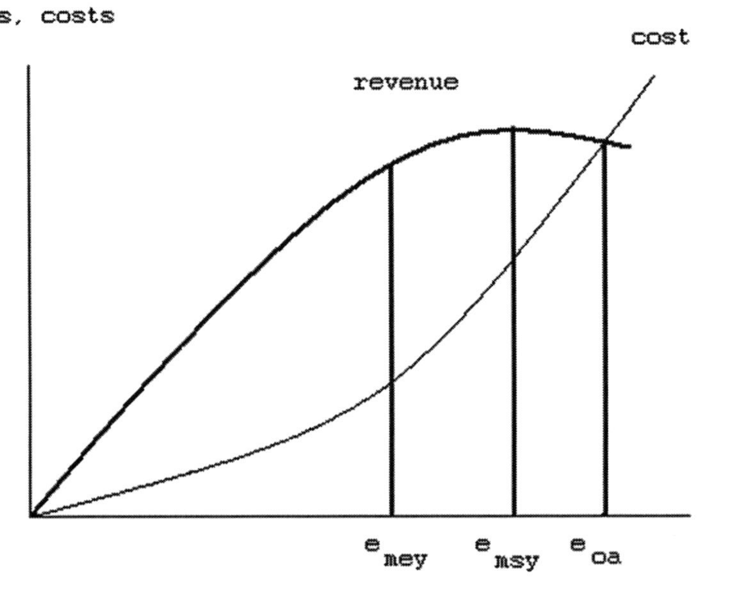

Three different effort levels are shown, each having an important economic meaning. The e_{oa} shows the effort level (theoretically) associated with open access. As long as there is rent to be extracted from the fishery (i.e. where revenue exceeds opportunity cost), there is an incentive for a new entry, and further entries will continue until all economic rent has dissipated. Rent is the surplus value that can be extracted from the fishing activity after all costs and normal returns have been subtracted from the revenues. In Figure 2.4, rent corresponds to the difference between the revenue and the cost functions. In the graph, new entrants lead to an increase in fishing effort, moving to the right on the graph until the effort level is at e_{oa}; at this point, revenues equal costs, there is no additional rent to be extracted, and there will therefore be no incentive for further entry into the fishery.

The e_{msy} is the level of effort where maximum physical yield, and therefore maximum revenue, is generated from the fishery. It is implicitly assumed in this graph that price is exogenous, and thus maximum revenue is achieved where catches are also at the maximum.

It is quite likely that the effort level e_{msy} is not the one that generates the highest rent which can be extracted from the resource. There are strong arguments for the economically optimal effort level to be lower, at least in a static setting. Rebuilding fisheries is a dynamic process and it is possible that the optimal effort level, from a purely economic point of view, is less than the effort level associated with MSY.

Box 2.3. Calculating MEY: An example from the Australian Commonwealth fishery

Kompas and Che (2008) have constructed a bio-economic model of selected stocks for the Commonwealth trawl sector of the southern and eastern scalefish and shark fishery. Solutions to the bio-economic model are obtained by maximising the aggregated discounted profits over time subject to a specification for harvest functions — the production function mapping fishing inputs to the harvest of fish — and the appropriate stock-recruitment relationship. All initial conditions for biomass are taken from virgin biomass measures provided by CSIRO or estimated from information supplied by CSIRO.

The results of the model are preliminary and the model likely requires further calibration based on biological studies and economic data. The results of the model are in two forms.

- Harvests and stocks in steady state (that is optimal harvests after stock rebuild).

- Harvests during the rebuild phase.

The preliminary results indicate that for four of the major stocks (orange roughy, pink ling, spotted warehou and tiger flathead) considerable stock rebuilding is required to maximise profits. That is, historical levels of harvest and fishing effort have resulted in current stock sizes that are below the stock level B_{MEY}. Also in the table below, the stock level associated with MEY relative to MSY is shown and for each species B_{MEY} is above B_{MSY}. The optimal harvests at the steady state are also shown in the table below. However, during the rebuild phase, harvests need to be set lower than 2004 catch levels to allow the stock to rebuild to B_{MEY}.

**Results of bio-economic model of the Australian Commonwealth trawl sector
of the southern and eastern scalefish and shark fishery**

Species	BMEY/ BCUR	BMEY/ BMSY	Optimal harvest at steady state (MEY) *tonnes*	Initial harvest TAC during rebuild * *tonnes*	Harvest (2004) *tonnes*
Orange roughy – Cascade	1.64	1.47	995	665	1 600
Spotted warehou	1.30	1.08	4 117	3 114	4 100
Pink ling (trawl)	1.80	1.29	1 397	914	1 073
Tiger flathead	1.05	1.03	3 830	2 980	3 200

** This is the initial Total allowable catch (TAC) during the rebuild phase. The TAC will increase through time over the rebuild period up to the optimal TAC at steady state.*

Source: Gooday *et al.* (2009).

Figure 2.4 depicts a common problem of fisheries management and does not in itself provide guidelines for how to rebuild fisheries. However, it does help make clear the distinction between biological targets (e_{msy}) and the socio-economic target (e_{mey}) Choosing the target is important. Although most rebuilding plans make reference to MSY as a target, there are obvious difficulties in using that particular target (Larkin *et al.*, 2007; Larkin *et al.*, 2011). In addition to the difficulties of calculating MSY for any particular stock , an obvious problem is that it is not clear how it can be used in fisheries where there are multiple species that interact. As long as species interaction comes into play, e.g. in the case of predator-prey relationships or where there is competition with regards to food or space, it becomes practically and theoretically impossible to maximise MSY for each and every species. Given that species interactions abound in the world's oceans it is questionable why MSY has been set as a reference point in the Johannesburg (WSSD) declaration.

From a purely economic viewpoint the measure that should be used as a target in fisheries management is the maximum economic yield (MEY). However this measure also has its strengths and weaknesses as it is often difficult to estimate and can be criticised for being too focused on monetary values.[6] Furthermore, a dynamic setting, where the aim is to maximise the net present value of the flow of MEY in the future, calls for an explicit social discount rate. Some progress has been made in implementing MEY in commercial fisheries (see Box 2.3, as well as Dichmont *et al.* (2009) and Larkin *et al.* (2007).

Rebuilding: How fast?

As rebuilding calls for an increase in biomass it is necessary to decrease fishing mortality at least temporarily as more fish must be left in the water to allow the stock to grow. If we assume, as in the model previously outlined, that there is a simple relationship between biomass and harvests then the question of how fast the target harvests should be achieved needs to be addressed.

Given simple assumptions about the relationship between biomass and harvest, Figure 2.5 shows three different harvesting trajectories which lead to the target harvest. The horizontal axis shows time, while the vertical axis shows the harvest. The three lines show different hypothetical scenarios of harvest levels over time. Implicit in each of the scenarios is that biomass will grow faster when the harvest rate is lower, so low harvest rates early in the rebuilding period will allow the target biomass, and therefore the target harvest rate, to be reached more quickly. All the rebuilding plans start at time zero (t_0). The thick line (scenario 1) shows a plan where a total moratorium is imposed until t_1 at which time the stock has recovered sufficiently to allow for the target harvest rate which in this scenario is then harvested immediately. The two other plans do not impose a moratorium but differ both according to the allowable harvest rates at the beginning of the plan and the adjustment of the harvest rate until the target harvest rate is reached. The thin line (scenario 2) presents a rebuilding plan where initially there is a relatively low harvest rate but due to a conservative (although not a moratorium) policy the harvest rate is allowed to grow relatively quickly until the target harvest rate (and target biomass) is reached at time t_2. The dotted line (scenario 3) shows a plan with a relatively high initial harvest rate but then a relatively slow increase in the harvest rate which means that the target harvest rate is not reached until at time t_3. Those three different scenarios can also be named according to the speed to which they reach the target biomass and harvest rate; "fast" (scenario 1), "medium" (scenario 2) and "slow" (scenario 3).

Although Figure 2.5 shows that a moratorium is the quickest way to attain the target biomass and harvest rate, one should be careful in drawing any conclusions on the optimal rebuilding strategy. Figure 2.5 shows hypothetical trajectories of harvest that lead to the target biomass and harvest rate given very simplistic assumptions. We have not yet discussed what the target biomass and harvest rate should be, nor the different costs and benefits of choosing one particular rebuilding plan. However the figure implicitly presents the concept of harvest or catch rules which are important in fisheries management and especially in the discussion of rebuilding fisheries. A harvest (fishing mortality) control rule is a function which shows what the level of harvest should be, given the stock size at any time. It follows that if the aim is to let the stock grow then the harvest control rule must lie below the sustainable catch curve. In other words, the harvest rate must be less than the growth rate at any given time. Such harvest rules are often referred to as catch-rules or feed-back rules in fisheries economics and have been widely studied (e.g. Anderson, 2010; OECD, 2009). An example of how the total allowable catch can be determined by a catch-rule is found in the Icelandic cod case study.

Figure 2.5. Three different harvest trajectories

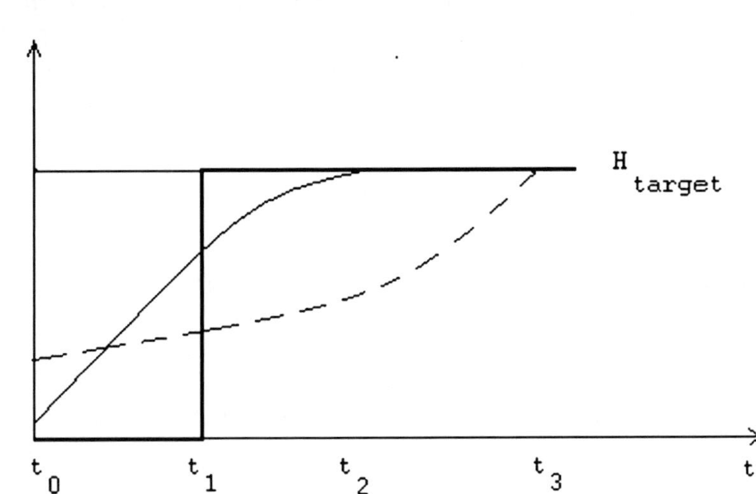

It is important to note that the actual target chosen for rebuilding is a policy choice. Whether the target for the rebuilding plan is MSY or MEY or any other target, the next thing to consider is how that target can be reached. In previous sections, the targets of rebuilding plans were examined without specifically indicating the rate at which these targets should be reached.

Both the speed of rebuilding and the target chosen depend to a large extent on the discount rate used. For rebuilding plans it seems plausible to use a social discount rate as opposed to a corporate discount rate, as rebuilding projects are akin to public investments. The discount rate should reflect how much future consumption or state of the world weighs against current consumption or state of the world. A higher discount rate places higher value on the present than the future and therefore discourages rebuilding efforts when compared to using a lower discount rate. In the same vein the discount rate may reflect how much the welfare of future generations should be taken into account by current generations. It also reflects uncertainties about the future as great uncertainties may encourage using higher discount rates. As rebuilding plans can be considered to be public projects the discount rate should reflect the opportunity cost of the public funds used for rebuilding, i.e. compared to different uses of funds. There is little consensus on the social discount rate to be used. Those who value the future equally to the present would argue for a zero discount rate while others would opt for a positive rate to reflect their preferences.

Disparity between public and private discount rates may create problems, especially with regards to gaining stakeholder support for rebuilding programmes. Although public authorities may see a rebuilding plan as yielding net benefits using a low social discount rate, private stakeholders may come to a different conclusion by using a higher private discount rate in their cost-benefit analysis, making future benefits less attractive compared to the up-front costs of rebuilding. In such cases, differences between the social and private discount rates may hinder a co-operative solution between the public and the private sector, an issue that surfaces in many other public investment projects. Possible solutions may include financial transfers from the public to the private sector to secure support for socially beneficial investments.

There is no single answer to how quickly or slowly a fishery should be rebuilt. The result of some modelling work which compares different time paths by using an "optimal" time path as a reference point is presented below. The analysis makes clear the differences in the time paths of rebuilding for different species that differ in biological characteristics, especially concerning growth rates, both in length and biomass. As expected, fast growing species have shorter optimal rebuilding times than slower growing ones.

From a pure bio-economic viewpoint the optimal harvesting rule is the one that generates the highest sustainable net present value of the flow of benefits from the resource over time. In reality both the choice of the target and how to get there (the rebuilding path) are policy issues. Policy makers may well choose to put the emphasis on other factors, for example by giving more weight to regional policies, enhancing biodiversity or distributional issues. However there are strong arguments to be made in defence of the bio-economic approach to the choice of the targets and the paths towards them. Some strengths and weaknesses of bio-economic analysis are presented below.

First, when estimating the net revenue flow it must be kept in mind that the net benefit flow should include all benefits deriving from various sources, such as monetary gain, existence values, etc., as well as all social costs, including direct monetary costs and non-monetary costs (or costs which are difficult to express in monetary terms) such as the cost of effects on the environment and biodiversity. The flow of net benefits should therefore reflect the true social value that is derived from the fishery.

Second, the choice of rebuilding path is a choice about moving benefits and costs between different time periods. For example, a sharp reduction in catches now with increased catches later, as opposed to a small reduction now with lower catches later means that benefits from the fishery are being transferred through time. It is important to note that, especially during long rebuilding time paths, stakeholders may not be able to reap the benefits of the rebuilding effort later and it is quite possible that the composition of the stakeholder groups may change over the time span of the rebuilding plan. It is necessary to take such facts into consideration when rebuilding strategies are being designed, especially if stakeholders´ participation is necessary for the success of the plan and if financial markets cannot be used to solve this problem. Bio-economic approaches allow for exploration of this distribution of benefits and costs through time and may suggest solutions.

Third, any successful rebuilding strategy should provide net benefits for society as a whole. This does not guarantee that each stakeholder will benefit. It is possible that the cost of the rebuilding effort will be borne by some stakeholders who will not profit from the same plan. Depending on the structure of the particular analysis, the bio-economic approach may help to shed light on the distribution among stakeholders, and on possible solutions. For example, it is possible to use multi-attribute objective functions in bio-economic analysis to address this problem.[7]

Fourth, Munro (2010) points out that natural resources are capital assets from the point of view of society. Therefore rebuilding plans should be analysed as investment programmes. There are, however, various difficulties that arise when comparing different plans, in particular as the human and physical capital used in fisheries is relatively non-malleable, i.e. it is not easy for workers to move to other sectors or to move capital out of the fishery to other uses. This holds true for vessels and other machinery but also for human capital, which has often been invested in the fisheries accumulating know-how and specific skills that are not easily used elsewhere. If all capital used in fisheries was perfectly malleable then the choice between rebuilding plans would be made easier. In that case it would be beneficial to choose the plan which rebuilds the fishery in the fastest way possible, e.g. by imposing a moratorium, taking into consideration that consumption now is more valuable than consumption later (given a positive discount factor). Although relatively little attention has been given to the role of non-malleable capital and the choice of rebuilding plans in the economic literature, this has been clear for many policy makers. The US Magnuson-Stevens Reauthorisation Act can be taken as an example. According to this Act, fisheries which are in need of rebuilding should be rebuilt within a time span of ten years where possible, but allowing for a longer rebuilding period which is deemed to be reasonable given the characteristics of the fishery.

More work is needed on clarifying some issues regarding the economic theory of rebuilding fisheries the most important being; the design of optimal harvesting rules for rebuilding fisheries, taking into account the time aspect; the non-malleability of human and physical capital; and how uncertainty and lack of information should be taken into account.

The methodology for ecosystem management approaches to rebuilding needs to be developed further. Although the importance of such an approach is evident and several countries have taken steps in this direction, it still needs to be clarified how it should be implemented with regards to rebuilding fisheries. This merits a thorough study which falls outside the scope of this publication.

Potential economic impacts from fisheries rebuilding

There are various factors that should be taken into consideration when evaluating the potential or actual economic impacts from rebuilding fisheries. Fisheries which are not operating at or near maximum potential with regards to economic benefits present a foregone value that could be retrieved under better management. The benefits of rebuilding include higher value of catches and lower cost of harvesting. The size of those benefits depends on the specific characteristics of each fishery.

The total impact of the world's ocean fisheries on world economic output has recently been estimated to be approximately USD 225-240 billion per year, taking into account both direct and indirect impacts (Dyck and Sumaila, 2010). A study commissioned by the World Bank estimated that the annual rent dissipated from the world's fisheries is around USD 50 billion per year (World Bank, 2008), mostly because of poor governance. Another study (Sumaila and Suatoni, 2005) estimated the potential economic benefits from rebuilding 17 stocks in the US fisheries to have a net present value of approximately USD 373 million.

A recent study (IDDRA, 2010) has estimated that the potential resource rent from UK fisheries might be around ten times the rent currently generated. Another study (Salz *et al.*, 2010) simulates the recovery of stocks and the elimination of overcapacity of seven important EU fisheries. According to calculations in the baseline scenario, nominal net profit from those fisheries could be increased almost five-fold within a 15-year rebuilding plan. Over the same period, the fleet size would be reduced from around 7 400 vessels to 5 700 vessels. Consequently, the net profit per vessel would increase by 520%. Although such rebuilding of stocks and decrease in the number of vessels is costly, the estimations point to almost EUR 500 million in net present value of profits over the 15-year rebuilding period. The average discounted net profit per vessel would be almost two times higher over the 15-year period than the 2005-2007 average.

Economic impact of rebuilding: Results from several case studies

Numerous case studies have been undertaken to describe different rebuilding plans and assess their outcome. Below is a brief summary of some of the findings of several case studies commissioned for this study where the focus is on the economic impacts. Many case studies do not provide much information on detailed and quantifiable economic impacts of rebuilding for various reasons, including lack of data availability or that not enough time had elapsed to fully evaluate the results. Nevertheless, most provide important insights into the socio-economic effects of rebuilding programmes. More details and evaluations on issues other than the economic impact are given in the case studies (*www.oecd.org/fisheries*).

Three case studies from Korea illustrate the development of fisheries rebuilding in that country. They underline the use that can be made of additional instruments, such as area protection and stock enhancement, coupled with efforts to reduce fishing effort. These case studies include three species: the sailfin sandfish, swimming crab and the yellow croaker. The rebuilding plan for sailfin sandfish started in 2006. Since the adoption of the plan, sailfin sandfish catches have increased; this is partly due to controlling the fishing effort, protecting spawning grounds, and active stock enhancement programmes. These increased catches have been accompanied by increased catch values. Although the limited availability of data does not allow for a sophisticated analysis of the total economic effects of the rebuilding effort, simply looking at the increase in fishing revenue associated with increased catches implies an increase of KRW 1 914 million (approximately USD 1.6 million) over the period 2005-06.

The rebuilding plan for the swimming crab fishery also started in 2006. Catches increased about three-fold from 2006 to 2008, due to both favourable environmental changes and the rebuilding effort. Catch values increased more than three-fold from 2005 to 2008. Using average prices, this increase amounts to an increase in catch value of around KRW 108 560 million (approximately USD 94 million).

The rebuilding plan for yellow croaker started in 2007. This stock is shared with neighbouring countries. Various measures to enhance the stock and habitat, along with effort reduction, seasonal closures and gear restrictions have had positive effects. The results have been greatly increased catches and harvest values that have increased by 50%. Using average prices, this amounts to an increase in catch value of around KRW 38 208 million from 2005-08 (approximately USD 33 million). Profitability is nevertheless thought to be low.

Three case studies from Japan include snow crab in Kyoto Prefecture, sailfin sandfish in Akita Prefecture, and chum salmon in Hokkaido. Rebuilding plans for snow crabs have been implemented since 1983, and catches slowly increased until they reached a peak in 1999. Since 2002, catches have decreased due mostly to a decrease in the number of vessels and per-vessel effort. The total landing value of snow crab increased from JPY 212 million (approximately USD 2 million) in 1983 to JPY 493 million (approximately USD 4.6 million) in 1995. Due to the reduction of the number of vessels and increased total value of landings in Kyoto, the annual landing value of snow crab per vessel has consistently increased and the figure almost doubled during the implementation period of the recovery plan. This increase in value is partly due to higher prices during strong economic growth. In 1983, the annual snow crab landings per vessel was less than JPY 10 million (approximately USD 100 000), and it increased to over JPY 20 million (approximately USD 200 000) by the mid-1990s. This increase has provided a strong incentive for fishers to continue the recovery plan.

The sailfin sandfish fishery was closed from late 1992 to late 1995. Since then a catch allocation scheme using TAC has been implemented. This fishery has experienced considerable fluctuations. Since the three-year closure, the stock has recovered significantly. Market conditions have, however, led to lower prices and average fisheries' household income is significantly lower than that of other households. The amount of landings of sandfish in Akita increased from 71 metric tonnes in 1991 to 143 tonnes in 1995. The volume of landings has increased to over 2 000 metric tonnes in recent years. Ex-vessel price of sandfish in Akita jumped to JPY 3 053/Kg (approximately USD 33/Kg) immediately after the three-year closure. However, prices decreased in subsequent years reaching a low of JPY 204/Kg. As a result, the total value of sandfish landings in Akita peaked at JPY 1 billion (approximately USD 11 million) in 2003 and then fell to JPY 0.57 billion in 2008.

The rebuilding programme of sandfish in Akita yielded positive results to fishers for the ten-year period after the fishery closure. Recently, however, the fishermen in Akita Prefecture have not been able to fully realise the benefits of stock rehabilitation because local prices for sandfish have fallen due to the supply of large amounts of sandfish from other regions. Average revenue per fisher was JPY 0.5 million before the fishery cessation, and increased to JPY 3 million per fisher in recent years. This per capita revenue increase is partly due to a reduction in the number of fishers after the fishery was closed.

The rebuilding plan for chum salmon can be traced back 120 years. In the early 20th century, catches decreased considerably but recovered after the 1970s, not least due to technical advances in hatchery methods and better water quality. It is true that the rebuilding programme has brought a significant increase in the volume of salmon returns in Hokkaido. It was less than 10 million salmons before the mid-1970s, but increased to the level of 50 million adults in recent years. Annual revenue of coastal salmon set-net operators were at JPY 60 billion per net in 1980s. In the 2000s, the sales dropped to the level of JPY 40 billion per-net. In sum, economic outcomes of the rebuilding programme were remarkable in the 1980s, but are currently at a substantially lower level due, in part, to the drop in the unit price for salmon.

Economic impacts of rebuilding: Results from a bio-economic model

Additional insights of the possible monetary benefits from rebuilding are provided in Costello *et al.* (2012). This study highlights many of the issues involved and estimates the costs and benefits of such plans given numerous biological characteristics. It also examines the value of not going through the process of overexploitation and later rebuilding, i.e. the value of maintaining a healthy stock. An overview of this study is provided below.

The model and main outcomes

The bio-economic model used has three linked components, i.e. a *biological stock model*, representing the biological dynamics of the fishery, a *harvest model* that relates catch biomass to stock biomass, and a *profit model* which evaluates the monetary value of the harvest, annual net profit and calculates the net present value of the fish resource taking into consideration the specific fishing policy, discount rate and a time horizon.[8] Using numerical methods, it is possible to estimate a fishing effort policy function that maximises the net present value of the fishery. This policy is called the "optimal policy" which is used as a benchmark to compare different rebuilding strategies. The model was parameterised to 18 hypothetical fisheries with different biological, harvest and economic characteristics.[9]

Table 2.1. Values of rebuilding from a collapsed state (baseline) and economic gains*

Net present values per year (2008, '000 USD)

Species	Baseline	Additional value if optimal	Additional value if fast	Additional value if slow	Rebuilding time in years		
					Optimal	Fast	Slow
Subtropical small pelagic	38 705	64 236	41 953	64 025	8	7	9
Subtropical shrimp	391	23 908	17 283	23 262	4	2	4
Subtropical grouper	997	1 779	1 655	1 788	5	3	5
Cold temperate scallop	23 943	96 499	92 621	94 382	15	5	16
Cold temperate flounder	9 561	37 306	29 508	36 126	6	3	7
Subtropical wrasse	58	131	117	124	10	4	10
Subtropical snapper	1 812	2 887	1 656	2 835	8	7	8
Subtropical jack	650	2 526	2 308	2 523	8	4	8
Temperate hake	56 999	228 427	182 698	218 226	7	2	7
Tropical/suptropical lobster	9 000	24 602	18 257	23 565	6	2	6
Temperate rockfish	23	17	13	18	26	19	29
Suptropical sparid	208	601	579	573	22	6	29
Warm temperate snapper	449	1 580	1 453	1 576	17	6	18
Cold temperate sole	4 783	1 405	1 430	773	5	3	6
Temperate monkfish	30 219	134 929	128 859	133 815	19	3	28
Temperate filefish	1 242	2 815	2 812	2 689	12	4	18
Subtropical clam	36	3	-7	3	4	4	5
Temperate small pelagic	9 654	22 282	20 010	22 223	24	14	25

* From rebuilding given different rebuilding times (optimal, fast and slow scenarios)
Source: Adapted from Costello *et al.* (2012).

Three different rebuilding strategies are compared, i.e. "fast", "slow" and "optimal", with the baseline case in which the fishery is not rebuilt and remains in a collapsed state.[10] In all cases, the fishery begins in a collapsed state. In the "fast" scenario, the fishery is closed until the stock biomass reaches the set target. In the "optimal" scenario, the fishery is rebuilt by fishing according to the economic optimal policy until the stock biomass reaches the set target. In the "slow" scenario, the fishing effort exceeds the economic optimal policy by 20% for the time period it would have taken to rebuild. When that point is reached, the policy reverts to the economic optimal policy until the biomass reaches the target. These three scenarios are compared with the net present value of maintaining the fishery in a collapsed state. The main results of this modelling exercise are shown in Table 2.1. For each of the 18 species, the net present value of the fishery is shown comparing different rebuilding strategies, optimal, fast and slow, with the baseline case. It also shows how many years it takes to rebuild according to the strategy chosen.

Monetary gains of different rebuilding strategies

Although the biological and economic characteristics of the fisheries differ, the preliminary results show that much can be gained by rebuilding fisheries. Comparing absolute values is not useful when comparing different rebuilding plans, while relative values are indicative. For the 18 species, there was on average a 575% increase in value resulting from rebuilding the stock from a collapsed state using an economically optimal strategy. Removing an outlier (subtropical shrimp) still yields an average relative increase in value of 250%.

These results indicate that there are considerable forgone income/profits by not choosing the economically optimal policy. The optimal rebuilding strategy yields on average a 22% greater value added than the fast strategy and an 8% greater value added than the slow strategy. These results depend heavily on the assumptions given for each strategy. Furthermore, given the assumptions, the slow strategy yields generally greater value added than the fast strategy. This result hinges, however, on the exact assumptions used and may change with different discount rates.

Time horizons for rebuilding fisheries

The time horizons vary greatly between fisheries and strategies. Using the optimal rebuilding strategy, stock recovery takes between 4 and 26 years (mean of 11 years). Choosing the slow strategy can mean that rebuilding may take between 4 and 29 years, depending on the species. This range is much less when using the fast strategy, where the shortest rebuilding period is only 2 years while the longest is 19 years.

The implications of different discount rates

It is worth noting the effect of different discount rates on the model's results. The results shown in Table 2.1 were derived using a discount rate of 1%, but at higher discount rates it is not economically optimal to rebuild some species as this implies a low value for future returns as compared to current benefits. As rebuilding is usually considered to be a long-term social investment benefiting future generations, public discount rates may be appropriate when calculating net present value of such investments. Also, various uncertainties regarding broader implications of not rebuilding, such as ecosystem considerations not directly accounted for in the analysis, may support the use of lower discount rates from a policy decision perspective. There is, however, little consensus on what the proper discount rate should be for public investment decisions.[11]

An interesting question is at which discount rate it becomes optimal not to rebuild the different species. This tipping point was found to be on average at a discount rate of 6% with a standard deviation of 2.6% for the modelled fisheries. This means that between 44% and 72% of the fisheries modelled are worth rebuilding at a discount rate of between 5% and 7%; between 78% and 100% of the modelled fisheries are worth rebuilding at discount rate between 2% to 3%.[12]

How biological characteristics affect rebuilding

With this model it is also possible to examine how biological characteristics correlate with the optimal rebuilding times. As expected, those species that grow quickly have shorter rebuilding times. Stocks that have low natural mortality rates have longer optimal rebuilding times, most likely because the natural turnover of biomass in the population is low. Interestingly, where the minimum legal size of the fish caught is large fisheries have shorter optimal rebuilding times, probably because more of the mature fish are protected from fishing and thus have more offspring to replenish the stock.

Correlation between different biological characteristics and relative monetary values of rebuilding also yields interesting insights. Although there is a positive correlation between the growth rate in length and the relative value of rebuilding, which reflects the fact that faster growing species recover more quickly, the correlation between growth rate in biomass and the value of rebuilding is negative. This result holds when both length and biomass growth parameters are taken into consideration. Accordingly, it is more profitable to rebuild species that grow faster in length than in biomass.

Generality of the results

Although this modelling exercise does not look at all the important aspects of rebuilding plans, e.g. the cost of decommissioning schemes, it provides important insights into the possible gains of different rebuilding strategies for a wide range of fisheries. Although the species chosen represent diverse life histories they are based on a geographically restricted selection from the United States and Mexico, mostly due to availability of data. In order to overcome this limitation, a global database of four key life history parameters for commercial marine species were run by the model. Comparison with the 18 species clearly demonstrates they are largely representative of commercial fisheries worldwide, which indicates that the main qualitative results hold for a wider range of species than the 18 chosen for this exercise.

These conclusions may be debated and hinge upon various assumptions. They nevertheless provide a starting point for creating hypothesis of the dynamics and value of different rebuilding strategies that can be tested, and clearly show that much can be gained by rebuilding fisheries.

Additional considerations about the potential impacts of rebuilding

From a socio-economic viewpoint the rebuilding of fisheries should result in additional benefits to society. However, how those benefits are distributed is a matter of concern for policy makers. In many cases, this is both a technical and a political issue and discussed further below.

Fisheries are never isolated from the rest of the economy or society. The process from the resource to final consumption or even discard of waste is a long chain of various phases and involves various stakeholders and economic considerations (Figure 2.6). It is difficult to assess the specific effects of rebuilding on each and every element of the value chain, but the objective should be to rebuild fisheries and thereby increase welfare to society as a whole. For this reason, a holistic approach, although often difficult to implement given limited knowledge and data, should be used in the rebuilding of fisheries.

There are other aspects of fisheries rebuilding, and fisheries management in general which are not captured by market forces as a result of a lack of markets. Examples include ecological considerations and existence values for species. Biodiversity is another example of such aspects that must be taken into consideration. In such cases government intervention to address such externalities is necessary if these factors are to be given the consideration they merit.

> ### Box 2.3. Studies on the economics of rebuilding
>
> There are several studies that take economic considerations that are explicitly taken into account in the analysis of rebuilding fisheries. The focus differs in each study; e.g. Hanna (2009) analyses various distributional issues, Munro (2009) looks at the role of incentives, Anderson (2009) discusses the technical issue of setting catch levels in rebuilding programmes, and Holland (2010) looks at how rebuilding strategies can be evaluated and compared.
>
> Other studies, like those of Worm *et al.* (2009), Sumaila *et al.* (2006) and the World Bank (2008), estimate the actual situation and the underlying problems which demand rebuilding efforts.
>
> Economic analysis suggests that rebuilding fisheries requires taking into account various factors such as the institutional structure and incentives, biological characteristics, and socio-economic aspects concerning the fishery in question.
>
> Most of these studies underline the dynamics inherent in the rebuilding process and how the situation may vary, not only from one area to another but also by species characteristics, such as life span, growth parameters, pelagic or demersal. Costello *et al.* (2012) provide a model taking different biological characteristics into consideration in a bio-economic model.
>
> Being a dynamic process, the choice of an appropriate discount rate for comparing costs and benefits which occur not at the same time is important (Azar, 2009; Costello *et al.*, 2012). The discount rate should reflect how society compares future benefits with current costs and must be decided by fisheries managers when evaluating and comparing different rebuilding strategies.

Figure 2.6. General effects of rebuilding

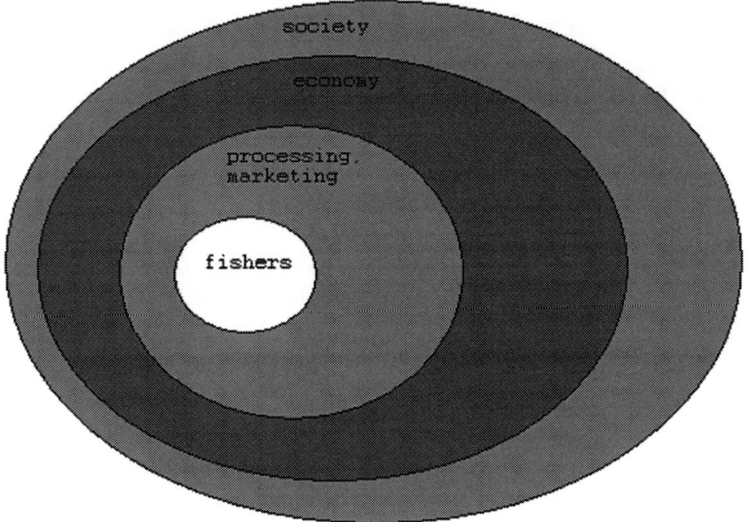

Uncertainty in rebuilding plans

Uncertainties affect fisheries management and rebuilding plans in many ways. These uncertainties reflect not only lack of knowledge concerning the biosphere, but also on the workings of the fishing activity itself and how this is affected by changes in natural and/or manmade conditions.

Uncertainties can be categorised by either type or source.[13] The type relates to where the uncertainty comes into play, e.g. economic, biological or political.

The main sources of uncertainties in the design and implementation of rebuilding plans can be classified as follows.

– *Process:* Uncertainty is due to random or chaotic processes arising from natural variability. An example is variability in recruitment over time, which is an important factor when designing a rebuilding plan.

– *Observation:* Uncertainty arises as a result of measurement and sampling errors. It is, for example, common that landing data suffers from such errors.

– *Model:* Uncertainties arise from the use of models. For example, the models may simply be wrong as the; assumed relationships can be either too simplistic or too complicated to be useful for decision-making.

– *Estimation:* Uncertainty relates to the fact that various parameters of the models must be estimated based on incomplete data. This source of uncertainty is well known in all fisheries models where collection of data is often difficult and costly.

– *Institutional*: Uncertainty relates to the uncertainties linked to the process of defining an effective plan. Institutional uncertainty can, for example, arise from difficulties in proper risk communication, or from institutional or legal issues due to the role of different stakeholders in the whole rebuilding process from design to implementation. Institutional uncertainty can also arise because of the lack of well-defined objectives leading to stated objectives which are in many cases not operationally feasible (Stephenson and Lane, 1995).

– *Implementation:* Uncertainty arises because it is not certain that policies will be successfully implemented. This may be due to many factors such as lack of institutional capacity, misaligned incentives, ineffective monitoring, and weak enforcement processes.

Although we have classified uncertainties according to their source or type they should not be looked at in isolation. Ludwig *et al.* (1993) have provided examples of how mixes of political, economic and biological uncertainties have led to a worsening of the fisheries situation (Box 2.4).

All of these uncertainties are present in fisheries management. As such, model uncertainty, implementation uncertainty and institutional uncertainty deserve special attention.

> ### Box 2.4. Various uncertainties and outcomes
>
> A classic example of how economic, political and biological uncertainties can interplay and lead to a worse situation of the fisheries is given by Ludwig et al (1993). The driving forces are described as economic and political incentives (forces). The effect, labelled a ratchet effect, works as follows: Given natural fluctuations in the stock size, additional investment will be made in "good" years. However, when the stock decreases to a size smaller than "normal" size, the industry appeals to government for help. The response is subsidies (direct or indirect). The effect is to encourage overharvesting. The ratchet effect is that no (or insufficient) limits are put on harvest investment during high stock levels, but political pressure not to disinvest during low stock levels are added. This reasoning has been used by Hennesey and Healey (2000) to explain the collapse of the stocks of the principal ground fish species off New England.
>
> *Source*: Brandt and Vestergaard (2011).

Model uncertainty and rebuilding

The fisheries environment and the economy are inherently complicated systems which are modelled using simplified assumptions. Such simplifications are necessary to keep the models tractable, operational and informative, but at the same time may create uncertainties with regards to the usefulness of the results and predictions. As an example, models often assume that relationships between different variables are linear or non-linear in a simplistic fashion and that changes are reversible. This is not always the case. Often a system pushed beyond a threshold stabilises in a new state from which it is not possible to revert to the former original situation. It is possible that in some cases the stock is already depleted beyond a threshold level where the growth rate becomes negative.[14] In that case, even removing fishing pressures completely will not allow the stock to grow. Another example is where markets, once lost, may be difficult to revive due to the arrival of substitute products.

Such complexities pose challenges which are best dealt with in the same way as other types of uncertainty, i.e. by using robust and adaptive models and decision mechanisms when designing rebuilding plans. These complexities should also be incorporated into the models used to develop the rebuilding plans, to the degree possible.

Institutional and implementation uncertainties in rebuilding

To counter the negative effects of institutional uncertainties it is important to consider risk and uncertainty issues when setting rebuilding objectives and adequately communicate options and results. No matter how well planned a rebuilding strategy is, reality is sufficiently complex that there will always be risks that rebuilding objectives are not met. Hence, attempts should be made to quantify risks and uncertainty so as not to create unrealistic expectations and to clearly indicate the tradeoffs being made through the rebuilding process.

Implementation uncertainty may hinder successful rebuilding even though fishing mortality is reduced, good management practices are introduced, and other favourable measures are implemented. A rebuilding plan for Irish cod illustrates this point. While many management measures were well implemented, the lack of communication about the associated risks led to significant frustration among fishers and fishery managers alike, which in turn jeopardised the entire rebuilding plan. As such, these researchers recommended improving the Irish rebuilding plan by including "clear, measurable performance targets, underpinned by sufficient data collection to assess performance of rebuilding, and an understanding of the inherent uncertainty involved." Scientists and economists should also communicate clearly the uncertainty and levels of risks involved in any rebuilding strategy (Kelly *et al.*, 2006).

Communicating uncertainty is a balancing act. Understating risk associated with a rebuilding plan may create criticism among stakeholders if the plan proves to be riskier than presented and could create the risk that stakeholders withdraw their support. If the uncertainties are overemphasised, however, stakeholder buy-in may be difficult to obtain in the first place. Therefore, estimates concerning uncertainty of rebuilding plans, and their assumptions, should be presented as accurately as possible and carefully communicated and discussed.

The potential use of the Management Strategy Evaluation framework

A formal risk analysis should be undertaken for each rebuilding plan where sources and different types of risk are analysed. The Management Strategy Evaluation (MSE) framework can be useful to identify and implement strategies for rebuilding that are robust to several types of uncertainty and are capable of balancing multiple economic, social and biological objectives.[15]

MSE is a general framework for designing and testing management procedures which in most cases specify decision rules for setting and adjusting TACs or effort levels to achieve a set of fishery management objectives.[16] An important feature of the framework is that simulation testing is used to determine how robust different management procedures are to uncertainty. Management procedures are usually selected so that there is a reasonable likelihood that a pre-specified and quantified management objective can be reached. MSE differs from simple harvest control rules in that the management procedures must specify the data and assessment methods used to link decisions to outcomes; for example, how the TAC that achieves the target fishing mortality rate is actually calculated.

A MSE framework usually incorporates a number of interlinked model components such as population dynamics, data collection, data analysis and stock assessment, a harvest control rule that specifies a management action, a harvest decision process, and an implementation plan for management. An operating model is then used to generate ecosystem dynamics including natural variations in the system. Data from the operating model are collected to mimic the collection of data from the fishery and their variability. These data are then fed into the assessment model. The outcome of the assessment model and the harvest control rule determines the management action. Fleet effort and catch are then modelled, taking into consideration potential errors in implementation and the resulting catches are fed back into the operating model. This cycle is then repeated to model the whole management cycle.

These interlinked model components allows for testing the effect of modifying different parts, such as by changing the operating model, as well as to test different assumptions about stochastic variability, etc. This also allows for testing alternative management scenarios by running numerous stochastic simulations over several years to see how well different procedures perform given different assumptions. Different management procedures can then be compared by how well they reach pre-determined objectives given the constraints. For example, one might look for a rule that leads to a low probability of stock collapse (e.g. a specific percentage of the simulation runs), has a low average variance in TACs and a relatively high average catch size. The choice of management procedures usually involves a compromise between different objectives which are often at odds.

MSE and the use of pre-specified management procedures to determine management actions has several potential advantages over the more common approach of using regular or periodic stock assessments followed by decisions on TACs. The MSE approach explicitly identifies the management procedures that are robust to variations, uncertainties and errors, both in the biological part of the model and its implementation. If done correctly, it leads to an explicit definition of management objectives that can be weighed against each other. As MSEs typically report a variety of indicators, this gives stakeholders the opportunity to consider the different trade-offs.

The MSE framework has its drawbacks. It is time-consuming and can reduce the flexibility of managers after implementation (Butterworth, 2007). This framework is also only as good as the underlying models and assumptions it relies on. Perhaps more importantly, the MSE framework has generally been developed without taking socioeconomic aspects into consideration. To become a useful tool for fisheries managers, the MSE framework should incorporate bio-economic models.

Additional considerations concerning uncertainty

Given the different types and sources of uncertainties it is tempting to look for general approaches to deal with uncertainty in the design of rebuilding plans. One way forward, proposed by Charles (1998), is to design the plans in such a way that they are *robust, adaptive and precautionary*. The overall objective should be to have the plan provide acceptable results even though our understanding of the fishery system itself is not complete.

The plans should be *robust*, in the sense that even though our knowledge is less than perfect, the plan will at least provide some level of success. This means that fisheries managers should prefer plans that perform well within the expected range of uncertainty.

The plan should also be *adaptive* in the sense that new information is taken into account. This calls for the plan to be flexible enough to make use of new information and knowledge. Incorporating input from various stakeholders may help to make management more adaptive to various changes during the fishing season.

Having a robust and adaptive rebuilding plan does not free fisheries managers from the problems of uncertainty. Therefore a *precautionary* approach is useful when balancing risks, e.g. between stock depletion and possibly foregone economic profits (Box 2.5). Under the precautionary approach, more uncertainty should be reflected in more conservative measures, e.g. in setting lower catch targets.

Box 2.5. The precautionary approach

The precautionary approach to fisheries management postulates that uncertainty should be taken explicitly into account by setting specific reference points which trigger specific actions. It further stipulates that the absence of scientific information should not result in lack of conservation actions. This approach requires that, given uncertainties, conservative actions are taken first and relaxed only when scientific evidence convincingly demonstrates that those actions are no longer needed. One can say that uncertainty favours the ecosystem, as opposed to harvesting. Seen in this light, the precautionary approach gives priority to preventing a crisis rather than responding to it (Garcia, 1994).

The precautionary approach to fisheries management is prevalent in many international agreements, such as the FAO Code of Conduct for Responsible Fisheries (FAO, 1995) and the UN Agreement on Straddling and High Migration Fish Stocks (UN, 1995).

Instruments for rebuilding

This section identifies the instruments available to fisheries managers and how they might be used. Once the objectives and trajectory of the plan have been set there is the question of "how to get there", or the choice of tools and policies that create the appropriate incentives to reach the targets. This is no one solution that will work in all situations as the particular approach selected will depend on objectives of management, knowledge of stocks, nature and type(s) of participants, the ability to monitor and enforce regulations, and stakeholders' involvement in the management process.

Table 2.2. Typology of management instruments

Control method	Control variable	
	Fishing effort (input control)	**Catch (output control)**
Regulatory (administrative technical measures)	• Mesh size • Size/amount of gear • Area/time closures	• Size and sex selectivity • TAC
Regulatory (administrative access control)	• Limitated[1] non-transferable[3] permits/licences (LL) • Individual non-transferable effort quotas (IE) • Territorial Use Rights in Fisheries (TURF) • Other types of effort limits	• Individual[2] non transferable[3] quotas (IQ) • Community-based catch quotas (CQ) • Other types of catch limits (maximum landings or vessel catch limits – VC)
Economic market-based (economic access control or "rights-based method")	• Transferable[3] licences[1] (LTL) • Individual transferable effort quotas (ITE)	• Individual[2] transferable[3] quotas (ITQ)
Economic not market-based	• Input[4] tax • Subsidy • Charges	• Landing tax • Subsidy • Charges

1. System restricting the number of vessels authorised to fish, their individual fishing capacity, and fishing time.
2. Individual quota = fraction of a Total Allowable Catch (TAC) allocated to a vessel or fishing firm.
3. Transferable = tradable on the market.
4. Components of fishing effort (intermediate consumption, fixed capital, labour).

Source: OECD (2006).

Fisheries managers have used different management instruments to try to safeguard stocks, enhance survival and reproduction of fish, and impede overfishing. Fisheries managers have also used various instruments to achieve economic targets such as lowering costs and increasing the value of catch. The OECD has published reports where the pros and cons of the main management instruments are described (OECD, 1997; OECD, 2006). The following discussion borrows heavily from these publications. Table 2.2 gives an overview of different management instruments used in various fisheries in OECD countries.

Classification of management instruments

Although management instruments vary, their aim is usually the same, i.e. to maintain productive fish stocks. Usually they are not specially designed to rebuild fish stocks or fisheries, but to control fishing mortality, either directly or indirectly, which is what is most often needed in rebuilding plans. It should be noted that in most cases fisheries managers use a mix of instruments in any given fishery (Box 2.6).

There are several possibilities for classifying management instruments. According to Table 2.2, management instruments are classified either by the control method used or what is actually being controlled (the control variable). Different control methods can further be classified as regulatory controls through technical measures, regulatory access controls, economic market-based controls, or economic non-market based controls. Control variables can be classified as input controls or output controls.

To make the discussion tractable, we classify different management instruments by the control variable, i.e. whether they are input controls or output controls.

The Northwest Atlantic Fisheries Organization (NAFO) managed Greenland halibut fishery rebuilding programme uses a mix of measures. According to this plan, the TAC in any year will not be more than 15% larger or smaller than the TAC of the preceding year. The contracting nations fish their respective shares under their national regimes, but NAFO requires that all vessels 24 metres in length or greater shall be subject to special licenses, and that the list of those vessels shall be available to NAFO. Additionally, authorised vessels may only land their Greenland halibut catches in ports designated by NAFO and these vessels are subject to inspection in port.

Box 2.6. Technical controls coupled with other measures

Usually technical measures (input and/or output) are not used in isolation but rather coupled with several other measures in rebuilding plans. Examples are found in the country case studies (*www.oecd.org/fisheries*).

A mix of measures are used in the Korean sailfin sandfish fishery rebuilding programme, such as licenses, spawning protected areas, nursing protected areas, limits on size of catch and mesh-size regulations. There are limits on gear size and the number of nets per boat. TAC is also set and the emphasis is on stakeholder participation and self-imposed management.

Input controls

Input controls restrict certain inputs used by fishers for their fishing activity. This restriction is usually implemented by direct legislation or regulation. The most common types of input restrictions are: limited licenses; gear restrictions, such as those regarding type and size of mesh in fishing nets; technical restrictions regarding vessel size; engine size; and time/area closures.

Although such measures have been widely used for decades, both theory and experience have shown that except under special circumstances they are inefficient in fisheries management, especially when used in isolation. Theoretically, the reason for this inefficiency of input restrictions in controlling fishing mortality lies in the simple fact that fishing effort is a mix of many factors of production, such as vessels, engines, fishing gear, and crew, and restricting the use of some inputs usually means that other inputs will be used more intensively as there is usually some substitution between the different inputs. Technical restrictions on inputs are also often difficult and costly to monitor.

Experience has also shown that input controls are rarely efficient. *Toward Sustainable Fisheries* (OECD, 1997) compiled case studies from around the world where this message was clear. There are cases of input restrictions being relatively successful in preserving the fish stock, such as the case of the Pacific halibut fishery in Canada, but that came at a high cost, such as waste of resource rent due to an inefficient race for fish and unstable supply to markets resulting in low prices and other waste (Munro, 2010).

The apparent inefficiency of technical input restriction highlights the important distinction that should be made between fishing effort and fishing mortality. When rebuilding a fishery calls for lowering of the fishing mortality it is not always sufficient to constrain one or several inputs of the fishing effort.

Regulatory access control for inputs

Rather than controlling the inputs used in a fishery directly, it is possible for fisheries managers to use instruments which limit access to the fishery and thereby seek to control real fishing effort and fishing mortality. Common instruments of this type are limited and include non-transferable licences, individual non-transferable effort quotas, and territorial use rights in fisheries (TURFs).

In most developed countries, fishers are required to hold a fishing license. It is possible to control effort to some extent by restricting the number of licenses under specific circumstances, although experience has shown that this is extremely difficult in most cases. The main reasons for this are that it is often difficult to exclude would-be fishers from acquiring licenses and a limit on the number of fishing licenses does not on its own directly control the fishing effort.

The problem with non-transferable licences and non-transferable effort quotas is similar to that of technical input controls. As real effort (i.e. the effort that determines fishing mortality) is difficult to measure there are problems with the efficiency of such measures. The fact that they are non-transferable adds to their inefficiency because although they limit the actual number of fishers or boats in a fishery, non-transferability makes it more difficult for a fisher or a vessel to leave the fishery. Although from a purely bio-economic perspective it might be seen as beneficial to reduce the number of fishers or vessels in the fishery, such limitations on trade are often set because of other aims, such as hindering concentration in the industry or as means to achieve certain aims of regional policy.

Individual non-transferable effort quotas give the holder of such a quota a specific quantity of effort units (input). Such effort units are usually denoted in some measure of fishing capacity and/or fishing time, such as number of allowable days of fishing, number of traps, nets or hooks on lines or in the number of fishing hours per day. As with other instruments that rely on input controls, they suffer from the fact that although controlling some input use, there will often be other inputs which are not controlled for and which become substitutes for the ones being controlled. Therefore, they may be difficult to use for controlling fishing mortality. The fact that such effort quotas are non-transferable makes the system rigid and it is more difficult for fishers or vessels to leave the fishery than if transferable.

Territorial use rights (TURFs) mean that a certain area of the ocean is allocated to a designated user or group, which then undertakes the fishing by allocating rights to users within the group. These use rights are usually coupled with a high degree of formal and informal transferability within the group. Territorial use rights have been used with some success in various fisheries. They are common in many fisheries where the targeted species are relatively sedentary and where the fishing area can easily be geographically delineated. Although TURFs have been used in various fisheries around the world, the most famous cases are in various Japanese inshore fisheries, such as the Sailfin Sandfishery in Akita Prefecture. Usually TURF-based management requires active stakeholder participation and has worked best where it is relatively easy to exclude other potential users from the fishing grounds and the resource.

Economic market-based input controls

Some of the inefficiencies of using limited-non transferable licensing and non-transferable effort quotas can be reduced by making them transferable. In that way, the market system for the buying, selling and leasing of those rights can be used to increase the efficiency of such instruments. If the number of fishing licenses is limited and they are transferable, they create exclusive rights to those who hold such a license. However, market-based input controls should not be considered as a panacea for fisheries management as they lead to inefficiencies due to substitution of factors of production, and are difficult and costly to monitor.

Economic non-market based instruments

For decades, fisheries managers have used several types of economic instruments which are not market based. Those instruments are used to control certain elements of the fishery, e.g. fishing effort, gear types, profitability and costs.

Such measures, e.g. subsidies, input tax, landing tax, have also been used to retrieve rent from the fishery. Taxes on landings are a common way to collect money to pay for necessary infrastructure, such as harbour facilities, and are thus more related to user-fees than tax levies.

Examples of such instruments aimed at reducing fishing effort are taxes on inputs, such as fuel or revenue taxes, as well as taxes on landings.

From the view of rebuilding fisheries, it is common to see such economic non-market instruments used to sustain a certain level of profitability for the fishers, for example as subsidies. Although they may temporarily ease the lives of the fishers, they have a negative effect on the resource and therefore a negative effect on sustainability and future benefits.

Although taxes can reduce fishing effort and may help in generating and collecting rent from fisheries, there are few examples of successful fisheries being managed through the use of taxes. The reason is most likely two-fold. First, it is technically difficult and requires much information to find the correct tax rate. In theory, the manager would have to know the cost functions of each and every fisher to do this in an optimal manner. Second, and probably more importantly, it is politically difficult to levy taxes on fishers where the fishery is in a crisis; conversely, this may explain why it is more common to see subsidies in fisheries.

Other types of economic non-market based instruments aimed at controlling inputs in fisheries are decommissioning and buy-back schemes, as well as training and educational programmes. A recent OECD study on decommissioning schemes provides guidelines on how fleet reduction methods can be implemented (OECD, 2009a). One of the most important lessons learned is that for a decommissioning scheme to be successful, it is necessary that fishers are restrained from reinvesting once they have been bought out of the fishery.

It has been mentioned above that human capital is an input in fisheries. Various countries have invested in training and educational programmes for those employed in fisheries, usually, but not always, with the aim to make them more efficient so as to raise their standard of living. From a rebuilding point of view, such programmes could also be used to diversify skills to make it easier for fishers to leave the fishery and engage in other employment, at least during the period of lower catches due to rebuilding.

Output controls

Regulatory technical output instruments

Output restrictions constrain the catch taken in a fishery. The most common output control is total allowable catch (TAC), which often is measured on the basis of landings. A TAC sets a maximum on the catch allowed for specific species, areas and time periods. It is among the most common management instruments used and is also used in combination with most other fisheries management schemes.

Although the setting of TACs is necessary for most fisheries management systems, it does not work well in generating benefits from a fishery if it is used in isolation. Theory and experience clearly shows that relying on TAC management alone results in overcapacity, shortened fishing seasons, and fluctuating landings (OECD, 1997). Additionally, over-exploitation has not generally been prevented in fisheries managed solely by TACs. Reasons for this may include the level of the TACs and lack of compliance.

It is nevertheless true that determining a TAC is a necessary precondition for a successful rebuilding plan. However, setting the appropriate TAC is not always straightforward especially when there is considerable uncertainty and different views about how to measure the stock and its carrying capacity. Transparency in the estimation of the TAC and information sharing with stakeholders on the process and assumptions behind the estimation is an important part of a fisheries rebuilding plan as a consensus or common understanding of the biological situation and the targets chosen can help ensure buy-in by relevant stakeholders. If fishers and other stakeholders disagree or are uncertain about the fundamental state of the stocks they may be less likely to adhere to and support a rebuilding plan. A consensus on the need to engage in rebuilding is a driving force for industry-initiated rebuilding plans as the case studies from Japan, Korea and Iceland clearly demonstrate.

In addition to total allowable catch, the size or sex distribution of the catch is often managed, usually through specification of allowable gear types and/or measures that restrict harvesting in certain areas or at certain times, e.g. to protect juveniles and strengthen the reproductive capacity of the stock.

Regulatory output controls

Regulatory output instruments restrict the amount of catch that individuals, vessels, companies or other groups may take from the resource. Such measures usually define rights for the participants in the fishery, although those rights may vary considerably in nature and form.

Instruments based on defining access rights to fisheries resources typically aim to remove the incentive to race for fish and to overcapitalise, thus improving the efficiency of fisheries resource allocation and use. These instruments have been or are being introduced into an increasing number of fisheries (e.g. OECD, 1997; OECD, 2006; EU, 2009). Examples include community-based quotas (CQs), individual quotas (IQs) and individual transferable quotas (ITQs).

It is commonly understood that overfishing arises from the fact that fishers impose negative externalities on each other because of the common property nature of fisheries resources. In other words, the lack of exclusive rights means there is a lack of incentives for fishers to exert the socially optimal amount of fishing effort. The problems arising from these negative externalities are not solved through market channels because there are no property or access rights in many fisheries. For this reason, several management systems seek to assign exclusive rights to fishers in order to internalise the negative externalities.

In some cases, such rights-based management systems are not initiated by government actions but through initiatives by fishers themselves (Ostrom, 1990). In other cases, government bodies initiate the setting up of such systems and allocate rights to fishers.

A brief account of some of the most common right- based management measures and their relation to fisheries rebuilding measures is given below.

Community-based catch quotas

It is possible to allocate catch quotas to a predefined fishing community which then decides on the allocation of rights within the community. The difference between community-based catch quotas (CQs) and TURFs is that the former are not applied to a specific geographical area. CQs are often used when formalising traditional access rights, e.g. in artisanal fisheries. Social cohesion and acceptance of the fisheries management plan is necessary for these to be able to support sustainable fisheries and generate benefits.

An interesting case is the snow crab fishery in Kyoto Prefecture in Japan. There has been a rebuilding plan in this fishery since 1983. The snow crab is a by-catch in the flounder fishery during the seasonal closing period. The rebuilding plan does not have specific targets but is based on technical and input controls, such as creating protected areas and closed seasons. Snow crab is a very important commercial species for the region and catches have been decreasing since reaching a peak in 1960. The stock has improved and in September 2008 became the first Japanese fishery to be certified under the Marine Stewardship Council (MSC) eco-labelling scheme. There are only 15 vessels which have a license for this fishery. Besides official regulations concerning, for example, TAC and restrictions on the number and size of vessels, there are also voluntary regulations including season length, closed areas, gear, and size limits. It is estimated that the major part of the stock recovery plan is based on these self-imposed regulations. Another interesting feature is the habitat enhancement work that has taken place. It is stressed in this case study that the members of the Kyoto Danish Seine Fishery Federation feel a strong ownership of the snow crab resource off Kyoto and that they have almost exclusive rights to access the resource. There is emphasis on lowering transaction costs and it is interesting to note that there are indirect access controls through the membership of traditional snow crab fishing families or groups. The result has been an increase in the value of landings per vessel, which again has reinforced the rebuilding plan.

Another interesting variation on this theme is the current Korean rebuilding programmes which also rely on active stakeholder participation and self-imposed management which are aimed at specific fisheries with exclusive rights to well-defined fishing communities (Lee *et al.*, 2006; Uchida *et al.*, 2010). Such self-imposed management is not solely confined to community management systems as the experience from the New Zealand hoki fishery clearly demonstrates. In that fishery, industry called for controls in addition to the setting of an appropriate TAC level.

The Mexican abalone fishery is an example of a fishery where widespread co-operation between different stakeholders can result in rebuilding action. Such bottom-up approaches require extensive co-ordination between stakeholders at different levels of governance. If successful, it may serve as a framework for other fisheries.

Similarly, the scallop fishery in St Brieuc Bay demonstrates how close co-operation between researchers and fishers can contribute to successful rebuilding in a co-management framework. The rebuilding of this scallop fishery has been successful in both economic and ecological terms.

The co-ordination within the group can be of many types, the most common being cleaning the fishing ground, monitoring illegal fishing, removing harmful species, and information exchange. Other forms of co-ordination include joint searching for good fishing grounds, restocking of targeted fish, assigning or rotating fishing grounds, as well as cooperation in developing and implementing various operational restrictions on factors such as size/age of catch, mesh size, amount of fishing gear, aggregate supply, duration of fishing operations, and designation of protected areas and seasonal closures. Some groups have also adopted quality control measures and co-ordinated marketing efforts. The government encourages the formation of such groups by transferring money to the groups to be used for "club activities."

Individual quotas (IQs)

Individual quotas restrict the catch of each fishing unit so that the sum of all quotas will be equal to the TAC. Theory and experience shows that individual quotas often lead to resource conservation as they eliminate the race to fish, improve safety, reduce gear conflicts and loss, lead to greater economic stability, and improve the quality of the fish landed.

Economic market-based output controls

Individual transferable quotas (ITQs)

The difference between individual quotas (IQs) and individual transferable quotas (ITQs) is that the latter are transferable through market transactions. Having the quotas transferable further lowers operating costs, improves resources rents and the investment climate, reduces fleet capacity, and increases profitability. But there might be social objectives that call for limitation on the transferability of quotas. The case of the

Norwegian Arctic cod fishery rebuilding plan shows how limitations on transferability can reduce the possible negative effect of quota trading on vulnerable regions and help to get stakeholder to buy into the rebuilding plan.

Comparing the New Zealand and Icelandic rebuilding experiences offers valuable insights. Both countries have used ITQ systems for rebuilding but differ with respect to implementation. According to the New Zealand Fisheries Act, all fisheries where the biomass is below the MSY level should be rebuilt. Although no specific time limits are set, the law stipulates that the TAC should be adjusted accordingly. In Iceland there is no such clause albeit a general one concerning all fish resources.

The experience from Canada's Pacific Commercial Groundfish Integration Programme also shows how ITQs can be used successfully as part of a rebuilding programme. An important part of this integrated plan was to issue ITQs for seven commercial fisheries, each of which uses different gears that target over 60 stocks along the entire Pacific coast of Canada. While the federal government provided the broad criteria and objectives for management, commercial fishers were empowered to develop a system that would attain these objectives. The implementation of ITQs for all species (including non-directed catch) combined with a rigorous monitoring system provided a compelling incentive for harvesters to curtail their catch of species undergoing rebuilding. However, through the flexibility afforded by the tradability of quotas, the integration programme also allowed harvesters to achieve economic efficiencies and thus maintain a high level of economic viability.

The experience from the Danish North Sea cod fishery shows how tradable property rights for catch quotas, gross tonnage, engine power, and days at sea have substantially decreased the number of vessels in the fleet. Although catches have decreased dramatically under the rebuilding plan, the landed value per vessel has decreased for some types of boats and gears while it has remained stable for others. Interestingly, economic projections indicate that an increased resource rent stems mainly from a reduction in fleet size rather than increased catches. This can be partly explained by the fact that increased catches of cod may lead to overexploitation of other species, which shows the importance of not taking a single-species approach to multi-species fisheries.

Experience shows that quota systems are efficient in aligning fishers' incentives to the objectives of rebuilding. In the Icelandic pelagic fisheries, fishers themselves called for a government intervention to avoid collapse. In the New Zealand hoki fishery, the industry itself called for additional controls other than simply lowering the TAC. In the Danish North Sea cod fishery, resistance from fishers toward stock recovery can partly be explained by the absence of future benefits from recovered stocks compared to short-term losses that they would face not only from the cod stock but also from other stocks that cannot be exploited fully or only at higher costs.

Using transferable quotas thus has several beneficial aspects but often at the cost of employment and increased concentration of quota ownership. For this reason, most countries that have introduced quota systems have at the same time set limitations on total quota holdings and rules regarding the transferability of quotas through time or between particular groups of quota holders. Furthermore, the initial allocation of quotas has proven to be an issue of concern in many cases.

Experience shows that for IQ and ITQ systems to function it is necessary to have good monitoring and surveillance systems, which are often costly.

Economic non-market output controls: Financial incentives

Various financial incentives, including taxes and charges, can be used to lower fishing mortality by restraining effort, while at the same time collecting rent from fisheries.

Taxes and charges can be used to restrain effort and collect rent from fisheries. An interesting case is the Mauritanian cephalopods fishery where economic non-market mechanisms have been used. In this specific fishery, there are input and output controls in use especially in the form of licenses both for foreign and national fleets. There are two groups of fishers, industrial and artisanal and care has to be taken to hinder clashes between the two, especially territorial conflicts. Small-scale fishers have *de facto* open access. Fishing agreements with foreign nations (especially the European Union) is an important source of income for the national treasury through the selling of fishing licenses. Nevertheless, the government has decided to introduce a new management plan as the ongoing management plan seems to be raising problems, especially with regards to redistribution issues.

Other indirect management measures

The management measures discussed so far are centred on fish stocks and fleets. There are, however, other management measures that take a wider view and are strongly linked to the development of fisheries management towards ecosystem management.[17] Two such measures are briefly examined below: enhancing habitat and enhancing stocks.

Enhancing habitat

Various spatial and area management techniques have been used in fisheries rebuilding strategies, ranging from marine protected areas (MPAs) to "no take" zones and area/time closures. Such tools can be designed to protect essential fish habitat or be structured to protect nursery and spawning grounds or other sensitive areas. However, fishing effort can be displaced to other areas or fisheries, and except in the case of sedentary species, area management zones are likely to encompass only some of the fish stocks. Nevertheless, marine reserves and MPAs may prove to be a complementary fisheries management tool to the traditional input/output controls.

Norway, for example, has introduced area based measures for fisheries management purposes in order to protect spawning grounds and vulnerable habitats, as well as to rebuild depleted stocks such as coastal cod, redfish and sand eel. The Japanese and Korean case studies show how habitat improvement can be implemented along with other measures to rebuild fish stocks and as an integrated part of a rebuilding plan.

Fisheries enhancement

Rebuilding fisheries through aquaculture-based techniques has been used with mixed results. Stocking may not be a solution for all fisheries, but could contribute to the rebuilding of coastal or sedentary species.

According to Bell *et al.* (2008) there are three types of enhancement related to fisheries.

– *Restocking* refers to the release of cultured juvenile fish into the wild in order to restore a severely depleted spawning biomass to a level where it can once again provide regular, substantial yields. This could also extend to the re-establishment of a species where it is locally extinct to rebuild a fishery or for conservations purposes (i.e. conservation hatcheries).

- *Stock enhancement* refers to the release of cultured juveniles into the wild to augment the natural supply of juveniles and optimise harvests by overcoming recruitment limitations.

- *Sea ranching* refers to the release of cultured juveniles into unenclosed marine and estuarine environments for harvest at a larger size in "put, grow, and take" operations. Note that the released animals are not expected to contribute to spawning biomass, although this can occur when harvest size exceeds size at first maturity or when not all the released animals are harvested.

To employ these techniques, adequate consideration must be given to the effect on the ecosystem as well as on the wild stocks, the economic benefits of taking such an approach, and how to integrate this with traditional fisheries management techniques.

The rebuilding plan for Chum salmon in Hokkaido, Japan incorporates stock enhancement. It is mainly based on setting annual catch limits for coastal set-net fisheries but there is a significant role played by hatcheries. Most of these hatcheries were previously in private hands and later nationalised, but some have been re-privatised in recent years. Non-governmental stakeholders play an important role in this rebuilding plan. Although no major economic analysis was conducted at the outset there is a report suggesting that the coastal set-net regulations were designed as self-imposed rules among fishers in order to minimise the transaction costs and curtail enforcement expenditures of the government (Kobayashi, 2009). This case shows the application of Japanese co-operative management. It has not been as successful as many other such management systems in Japan, not the least because of a more complicated stakeholder mix and the existence of additional complexities, such as the hatching activity.

The Japanese and Korean case studies illustrate that stock enhancement can be implemented as part of a broader suite of measures in rebuilding plans.

Additional observations regarding instruments to rebuild fisheries

There is no single answer as to what instruments are best suited to rebuild fisheries. The choice of instruments hinges upon many factors but there are some lessons to be learned from theory and experience.

First, simply setting a TAC for each and every species in a fishery is not enough as the underlying forces which lead to excessive harvesting and rent dissipation are still at play.

Second, rights-based fisheries management measures have proven to be effective in managing fisheries and there is every reason to believe that they are also effective in rebuilding fisheries given the right incentives (Grafton *et al.*, (2005), Sutinen, 1999; Larkin *et al.*, 2007). Under special circumstances, rights-based management systems might lead to the extinction of some species, which would usually run counter to ecosystem management objectives. However, such circumstances are not likely in most fisheries, although scholars do not agree on how unlikely they are (Grafton *et al.*, 2007). However, rights-based management systems based on output controls (quotas) have proven to be efficient in controlling exploitation, while generating rent and profits in fisheries and reducing the number of participants (Sutinen, 1999). With regards to rebuilding, the most important lesson is that rights-based management systems have proven to be effective in protecting fish stocks and habitat.

Third, technical input controls have proven to be inefficient in limiting fishing mortality. They should not be considered to be the first choice when rebuilding fisheries.

Fourth, a study by Sutinen (1999) on the effectiveness of different management instruments in OECD countries showed that time and area closures were not very effective in assuring resource conservation. They may, however, be necessary in rebuilding plans where the rebuilding of species is a part of the plan and it is deemed necessary to protect some subset of the population or its habitat, such as spawning grounds and/or spawning fish.

To this must be added that fisheries and countries differ considerably. In many cases it may simply not be possible to use specific types of instruments due to various factors.

When choosing which instruments to use fisheries managers must take into account issues such as data availability, monitoring and surveillance abilities, costs and benefits of different management instruments, cultural issues and traditions, and national and international legislation and instruments. Certain types of management measures can prove to be cost prohibitive and/or non-enforceable due to the lack of monitoring and surveillance as well as having to rely on data which is not available. Cultural issues and traditions may create opposition to an otherwise well designed rebuilding plan, while national and international law may block certain types of management actions.

It is clear that there is no single answer to the question of which instruments fisheries managers should use for rebuilding. However, it helps to know the limitations and virtues of different instruments and to compare them with the realities in which they are to function.

When faced with a fishery in the need of rebuilding, it is not likely that the management system that resulted in the fishery being in that state is the one that is best suited for the rebuilding effort. If the fishery's state is due to overfishing and/or rent dissipation, and not biological or environmental factors, then it is clear that changes are needed in the way the fishery is managed. How those changes can be brought about, and the many hindrances in that process, is the subject of the next chapter.

Notes

1. The Workshop on the Economics of Rebuilding Fisheries was convened by the OECD's Committee for Fisheries, and held in Newport, Rhode Island, United States on 21-22 May 2009. See *The Economics of Rebuilding Fisheries: Workshop Proceedings*, (OECD, 2010).

2. Maximum sustainable yield is defined as the largest average catch or yield that can continuously be taken from a stock under existing environmental conditions. See, for example, Parker (2003).

3. On this point, see Sutinen (2008).

4. For example, the MEFEPO project on ecosystem-based fisheries management (*www.liv.ac.uk/mefepo*).

5. On the ecosystem approach to fisheries, see FAO (2003).

6.   In theory, the MEY should include all relevant costs and prices, including environmental and social costs and benefits. However, valuation of many of these costs and benefits requires a great deal of information, much of which is rarely available.

7.   For a more detailed discussion on this point, see OECD (2010).

8.   The biological model is a discrete in time, non-spatial, size-structured population model with three size classes and Beverton-Holt type recruitment. The harvest model uses a linear relationship between catches and stocks using a catchability parameter. For further details see Costello *et al.* (2012).

9.   The example real-world fisheries and data sources that were used to develop these hypothetical fisheries are listed in Costello *et al.* (2012).

10.   The collapsed state is where stock biomass is reduced to 50% of its value a MSY or to as close to that level as possible.

11.   For a discussion, see Zhuang *et al.* (2007) and Azar (2009).

12.   These percentages refer to the number of modelled fisheries.

13.   For a discussion on the types of uncertainties, see Brandt and Vestergaard (2011).

14.   This is called dispensation in the biological literature and can occur due to various factors such as reduced probability of finding a mate or increased pray per offspring. See Liermann and Hilborn (2001) for a discussion.

15.   On the Management Strategy Evaluation Framework and its use in different fisheries, see Holland (2010) on which this discussion is largely based.

16.   There are very few examples of MSEs that have explicitly incorporated economics or economic objectives but incorporating bio-economic models into the MSE framework could provide management advice to fisheries managers and stakeholders. See Holland (2010).

17.   For a discussion on the ecosystem approach to fisheries management, see FAO (2003).

References

Andersen, P., J.L. Andersen and H. Frost (2010). ITQs in Denmark and Resource Rent Gains. *Marine Resource Economics*, 25, pp. 11-22.

Anderson, L.G. (2010). "Setting allowable catch levels within a stock rebuilding plan" in OECD Workshop Proceedings: Economics of Rebuilding Fisheries. Rhode Island.

Anderson, L.G. and J.C. Seijo (2010). *Bioeconomics of Fisheries Management*. Wiley-Blackwell.

Azar, S.A. (2009). A Social Discount Rate for the US. *International Research Journal of Finance and Economics*, Vol. 25.

Bell, J.D., K.M. Leber, H.L. Blankenship, N.r. Loneragan and R. Masuda (2008). A new era for restocking, stock enhancement and sea ranching of coastal fisheries resources. *Reviews in Fisheries Science*, 16, pp. 1-8.

Binet, T. (2010). *Cephalopods in Mauritania*, OECD internal document, Paris.

Brandt, U.S. and N. Vestergaard (2011). Assessing Risk and Uncertainty in Fisheries Rebuilding Plans, Working Papers 107/11, University of Southern Denmark, Department of Environmental and Business Economics.

Butterworth, D.S. (2007). Why a management procedure approach? Come positives and negatives. *ICES Journal of Marine Sciences*, 64.

Caddy, J.F. and D.J. Agnew (2004). An Overview of Recent Global Experience with Recovery Plans for Depleted Marine Resources and Suggested Guidelines for Recovery Planning, *Review of Fish and Fisheries*, Vol. 14, pp. 43-112.

Caddy, J.F., Mahon, R. (1995). *Reference points for fisheries management*. FAO Fisheries Technical Paper, No. 347, Rome.

Costello, C., B.P. Kinlan, S.E. Lester and S.D. Gaines (2012). The economic value of rebuilding fisheries. *OECD Food, Agriculture and Fisheries Working Paper N°55*, OECD, Paris.

Davis, J.C. (2010), *Rebuilding Fisheries: Challenges for Fisheries Managers* in OECD Workshop Proceedings: Economics of Rebuilding Fisheries. Paris.

Dichmont, C.M., S. Pascoe, T. Kompas, A.E. Punt, and R. Deng. (2010). On Implementing Maximum Economic Yield in Commercial Fisheries. *Proceedings of the National Academy of Sciences* (PNAS) 107(1).

Dick, A.J. and U.R. Sumaila (2010). Economic impact of ocean fish populations in the global fishery. *Journal of Bioeconomics*. Published online: 19 August 2010.

EU (2009) *Green Paper. Reform of the Common Fisheries Policy. Brussels. eur-lex.europa.eu/LexUriServ/LexUriServ.do?uri=com:2009:0163:fin:en:pdf*.

FAO (1995), *Code of conduct for responsible fisheries*. Food and Agriculture Organisation of the United Nations. Rome. FAO (2002), *CWP Handbook of Fishery Statistical Standards*. Section G: FISHING AREAS - GENERAL. CWP Data Collection. In: FAO Fisheries and Aquaculture Department [online]. Rome. Updated 10 January 2002. [Cited 12 December 2011]. *www.fao.org/fishery/cwp/handbook/G/en*

FAO (2006), *Stock Assessment for Fishery Management: A Framework Guide to the Stock Assessment Tools of the Fisheries Management Science Programme*, FAO Fisheries Technical paper No. 487, FAO, Rome.

Garcia, S.M. (1994). The precautionary principle: Its implications in capture fisheries management. *Ocean and Coastal Management*, 22, pp. 99-125.

Gooday, P., T. Kompas, NT. Che and R. Curtotti (2009) "Harvest Strategy Policy and Stock Rebuilding for Commonwealth Fisheries in Australia: Moving Toward MEY" in OECD Workshop Proceedings: Economics of Rebuilding Fisheries. Paris. Grafton, R. Q., T. Kompas, and R.W. Hilborn (2007), "Economics of Overexploitation Revisited", *Science*, Vol. 318. no. 5856, p. 1601.

Hanna, S. (2009). "Managing the transition: distributional issues of fish stock rebuilding" in OECD Workshop Proceedings: Economics of Rebuilding Fisheries.. Paris..

Hilborn, R. (2007). Managing fisheries is managing people: what has been learned? *Fish and Fisheries* 8, pp. 285–296.

Hobday, A.J., A. Smith, H. Webb, R. Daley, S. Wayte, C. Bulman, J. Dowdney, A. Williams, M. Sporcic, J. Dambacher, M. Fuller, T. Walker. (2007) Ecological Risk Assessment for the effects of Fishing: Methodology. Report R04/1072 for the Australian Fisheries Management Authority, Canberra.

Holland, D. S. (2010), "Management Strategy Evaluation and Management Procedures: Tools for Rebuilding and Sustaining Fisheries", *OECD Food, Agriculture and Fisheries Working Papers*, No. 25, OECD Publishing.

IDDRA (2010). The Potential Benefits of a Wealth-based Approach to Fisheries Management: An Assessment of the Potential Resource Rent from UK Fisheries. DEFRA Project MF 1210.

Jakobsson, J., G. Stefansson (1998), "Rational harvesting of the cod – capelin - shrimp complex in the Icelandic marine ecosystem," *Fisheries Research*, No. 37, pp. 7-21.

Kelly, Ciaran J., Edward. A. Codling, and E. Rogan (2006). "The Irish Sea cod recovery plan: some lessons learned", *ICES Journal of Marine Science*, Vol. 63, pp. 600–610.

Kobayashi, T. (2009). History of salmon stock enhancement in Japan. Hokkaido University Publication, Sapporo, Japan (in Japanese).

Kompas, T. and TN Che (2008). *Maximum Economic Yield in the Southern and Eastern Scalefish and Shark Fishery*, ABARE Report to the Fisheries Resources Research Fund, Canberra, February.

Larkin, S.L., G. Sylvia, M. Harte and K. Quigley (2007). "Optimal Rebuilding of Fish Stocks in Different Nations: Bioeconomic Lessons for Regulators", *Marine Resource Economics*, Vol. 21, pp. 395-413.

Larkin, S., S. Alvarez, G. Sylvia, and M. Harte. (2011), "Practical Considerations in Using Bioeconomic Modelling for Rebuilding Fisheries", *OECD Food, Agriculture and Fisheries Working Papers*, No. 38, OECD Publishing.

Lee, Sang-Go (2009). "Rebuilding fishery stocks in Korea: A national comprehensive approach", in OECD Workshop Proceedings: Economics of Rebuilding Fisheries. Paris.

Lee, K.N., J.M. Gates and J. Lee (2006). Recent Developments in Korean Fisheries Management. *Ocean and Coastal Management*, 49, pp. 355-366.

Liermann, M. and R. Hilborn (2001). Depensation: evidence, models and implications. *Fish and Fisheries*, Vol. 2.

Ludwig, D., R. Hilborn, C. Waters (1993). Uncertainty, Resource Exploitation, and Conservation: Lessons from History. *Science*, Vol. 260, 2 April.

Munro, G. (2010), *Getting the economics and the incentives right: instrument choices in rebuilding fisheries* in OECD Workshop Proceedings: Economics of Rebuilding Fisheries. Paris.

Myers, R.A., J.K. Baum, T.D. Shepherd, S.P. Powers, and C.H. Peterson (2007). "Cascading Effects of the Loss of Apex Predatory Sharks from a Coastal Ocean", *Science*, Vol. 315, no. 5820.

North, D.C. and P.T. Robert, *The Rise of the Western World: A New Economic History*, New York, Cambridge University Press, 1973.

OECD (1997). *Towards Sustainable Fisheries*. Paris.

OECD (2006). *Using Market Mechanisms to Manage Fisheries*. Paris.

OECD (2009a). *Reducing Fishing Capacity: Best Practices for Decommissioning Schemes*. Paris.

OECD (2009b). *Making Reform Happen in Environmental Policy*, OECD, Paris. (*ENV/EPOC/WPNEP(2009)4/FINAL*)

OECD (2009c). *Strengthening Regional Fisheries Management Organisations*. Paris.

OECD (2010). *The Economics of Rebuilding Fisheries. Workshop Proceedings*, Paris.

OECD (2011). *Fisheries Policy Reform. National Experiences*. Paris.

Ostrom, E. (1990), G*overning the Commons: The Evolution of Institutions for Collective Action*. Cambridge University Press.

Parker, S.P. (2003). McGraw-Hill Dictionary of Scientific & Technical Terms, 6th edition. McGraw-Hill.

Rosenberg, A.A. and C.B. Mogensen (2007), *A Template for the Development of Plans to Recover Overfished Stocks*, WWF, www.panda.org/marine.

Salz, P., E. Buisman, H. Frost, P. Accadia, R. Prellezo and K. Soma (2010). *Study on the remuneration of spawning stock biomass*. Final Report, Framian. *ec.europa.eu/fisheries/documentation/studies/remuneration_of_the_spawning_stock_biomass_ en.pdf*.

Sandberg, P. (2009). Rebuilding the stock of Norwegian spring spawning herring. Lessons learned, in OECD Workshop Proceedings: Economics of Rebuilding Fisheries. Paris.

Stephenson, R.L., and Lane, D.E. 1995. "Fisheries management science: a plea for conceptual change". *Canadian Journal of Fisheries and Aquatic Sciences*, 52, pp. 2051–2056.Sumaila, U.R., A. Khan, R. Watson, G. Munro, D. Zeller, N. Baron and D. Pauly (2007), "The World Trade Organization and global fisheries sustainability", *Fisheries Research* 88, pp. 1–4.

Sumaila, U.R. and E. Suatoni (2006). Economic Benefits of Rebuilding U.S. Ocean Fish Populations, *fisheries Centre Working Paper* No. 2006-04, University of British Columbia, Vancouver, B.C., available at *www.fisheries.ubc.ca/publications/working/index.php*

Sutinen, J.G. (1999). What works well and why: evidence from fishery management experiences in OECD countries. *ICES Journal of Marine Science*, Vol. 56.

Sutinen, J.G. (2008). Major Challenges for Fishery Policy Reform: A Political Economy Perspective, OECD Food, Agriculture and Fisheries Working Paper No. 8, available online at www.oecd.org/fisheries.

Swasey, J.H. and A.A. Rosenberg (2006), *An Evaluation of Rebuilding Plans of US Fisheries*, Lenfest Ocean Program, Washington DC, April.

Troadec and Boncoeur, J. (2003). Economic Instruments for Fisheries Management, unpublished report to the OECD.

Uchida, H. (2009). Community-based management for sustainable fishery: lessons from Japan. In *OECD Workshop Proceedings: Economics of Rebuilding Fisheries*, OECD, Paris.

Uchida, H., E. Uchida, J-S. Lee, J-G. Ryu and D-Y. Kim (2010). Does Self Management in Fisheries Enhance Profitability? Examination of Korea's Coastal Fisheries. *Marine Resource Economics*, Vol. 25.

UN (1995). *Agreement for the Implementation of the Provisions of the United Nations Convention on the Law of the Sea of 10 December 1982 Relating to the Conservation and Management of Straddling Fish Stocks and Highly Migratory Fish Stocks*, A/CONF.164/37.

UN (2002). *Report of the World Summit on Sustainable Development, Johannesburg, South Africa*, 26 August–4 September 2002, Chapter 1.2, Plan of implementation of the WSSD (*www.Johannesburgsummit.org*).World Summit of Sustainable Development website: *www.un.org/events/wssd/*

Yagi, N. (2010), *Stock Rebuilding Plan for Chum Salmon in Hokkaido, Japan*, OECD internal document, Paris.

Yamazaki, S., T. Kompas and R.Q. Grafton (2009), "Output versus Input Controls under Uncertainty: The Case of a Fishery" in *Natural Resource Modeling*, N°2.

Wakeford, R.C., D.J. Agnew and C.C. Mees (2007), *Review of Institutional Arrangements and Evaluation of Factors Associated with Successful stock Recovery Programs*, CEC 6th Framework Programme No. *022717 UNCOVER, MRAG Report, March.*

World Bank, (2008). *The Sunken Billions. The Economic Justification for Fisheries Reform.* Washington D.C.

Worm, B, R. Hilborn, JK Baum, T.A. Branch, J.S. Collie, C. Costello, M.J. Fogarty, E.A. Fulton, J.A. Hutchings, S. Jennings, O.P. Jensen, H.K. Lotze, P.M. Mace, T.R. McClanahan, C. Minto, S.R. Palumbi, A.M. Parma, D. Ricard, A.A. Rosenberg, R. Watson, D. Zeller (2009). "Rebuilding global fisheries", *Science* N°325, pp. 578-585.

Zhuang, J. Z. Liang, T. Lin and F. De Guzman (2007), "Theory and Practice in the Choice of Social Discount Rate for Cost-benefit Analysis: A Survey", *ERD Working Paper*, No. 94, Asian Development Bank. May.

Chapter 3.

Lessons learned from case studies to rebuild fisheries

These case studies seek to identify the factors underlying the outcomes – successful or not – of various rebuilding plans and efforts. They cover many different fisheries both at the national and international levels and there are a set of common lessons to be learned. These include the importance of integrating economics early in the rebuilding design process as various social and economic aspects may hinder or help in the execution of the plan. This also underlines the importance of stakeholder involvement in designing the plans. If stakeholders are strongly opposed, the chances of success are low. Incremental approaches can be helpful, especially in situations where this is much uncertainty and little reliable data. The case study material also shows that monitoring and enforcement are necessary in order to deliver successful outcomes. Rebuilding international fisheries calls for joint and co-ordinated efforts of all countries involved in the fishery.

A central component of the OECD rebuilding project was the collection of case studies of fisheries rebuilding plans at the national and international levels to identify the factors underlying the outcomes of rebuilding programmes and efforts.

This chapter underlines the main lessons from the case studies. It also presents a literature review of other initiatives that assess rebuilding plans and to provide an overview of the case study methodology for this project. Additional information and the case studies are available in the OECD *Food, Agriculture, and Fisheries Working Paper series* (*www.oecd.org/fisheries*).

The objective is to identify the critical aspects of fisheries rebuilding plans that are useful to policy and decision makers in the formulation and implementation of future rebuilding plans, or in the revision of existing ones. There is a particular focus on the economic and institutional factors that facilitate or impede fisheries rebuilding so as to complement recent studies that have primarily examined biological and management factors. It is also important to have a basic understanding of the biological factors associated with each rebuilding case, as stock characteristics and basic biological traits are a significant factor in the success of rebuilding. Indeed, biological aspects such as fecundity have a strong role to play in the timeframe for rebuilding (e.g. short lived species may require less time to rebuild, while long living, slow growing species generally require longer time horizons) and are central in determining the types of rebuilding measures and timeframes that may be most effective for a particular situation. The case studies are intended to bring useful insight regarding the key elements of rebuilding plans and provide a rich dataset from which to extract a set of considerations as the basis for a set of best practice guidelines.

The case study is a research tool that allows for a holistic, comprehensive review of a complex and multifaceted issue (Feagin, Orum, and Sjoberg, 1991), and is effective when a limited number of examples are examined with a reasonable amount of detail. According to Yin (2004), the case study approach is appropriate in circumstances where the research question is broadly defined, where "complex multivariate conditions" as opposed to "isolated variables" are involved, and evidence must be drawn from multiple sources.

The case study approach in support of the OECD project on the economics of rebuilding fisheries will provide a compliment to the quantitative assessment of other projects studying rebuilding fisheries (e.g. the World Bank Rent Drain project). Such evidence-based research provides the necessary implementation examples of successful and unsuccessful elements of rebuilding plans and support the proposed development of best practice guidelines.

The case studies have been undertaken in three ways: by OECD in co-operation with individual Member countries; by consultants; and by member countries.

The proposed analysis represents an assessment on the common elements of fisheries rebuilding plans from an international perspective. It provides a systematic examination of rebuilding plans in OECD and non OECD countries that is intended to yield useful insights for policy and decisions makers, fisheries managers and others involved in the development of rebuilding plans. It should be noted that it is not the intention to evaluate the success or failure of individual rebuilding plans. The level of success of a particular fisheries rebuilding plan may nevertheless be measured against any objective or milestone identified within the plan itself.

The primary purpose is to develop an enhanced understanding of the issues associated with the development and implementation of rebuilding plans and provide information across the range of approaches used in different countries. Ultimately, this work contributes to the development of a set of best practice guidelines regarding the design, implementation or modification of rebuilding plans in both OECD and non OECD countries.

The inclusion of cases from a wide diversity of countries, geographic areas, fish stock characteristics, and institutional structures allows for a robust assessment of the various issues that arise in the development and implementation of a disparate set of rebuilding plans.

Analytical framework

The case studies have been developed through the use of a template developed with a view to providing a consistent structure for analysis. The criteria and indicators in the template cover institutional and management arrangements, economic, social, environmental, and biological criteria, and help highlight the roles of each of these aspects in fisheries rebuilding.

Case study selection

The overarching selection was undertaken so as to target the various thematic issues and ensure that they are incorporated, to the extent possible, to allow for a rich and diverse set of case studies. In particular, the primary characteristics include (but are not limited to) the following factors.

–   *Vary the types of fisheries or industry groups.* This includes rebuilding fisheries in a multi-sector context which is composed of many fishers with disparate interests as compared to large consolidated industrial fleets; single stock or mixed stock fisheries as these cases may yield useful insights on the complexities of each situation. Further, case studies should span across the scale of the fisheries (coastal/inshore/deepwater) as each level may represent unique challenges.

–   *Management tools and approaches.* Consideration of various types of management regimes. Management responses and tools used to rebuild fisheries vary across plans depending on the cause of the fisheries depletion and could include some combination of the following: input/output controls, rights based tools, stock replenishment, and/or habitat enhancement. Given that each fishery has its own distinctive characteristics, the rebuilding case studies should reflect the various tools available, particularly as there is no one solution to fit all fisheries.

–   *Economic and social aspects:* Ideally, fisheries of various economic importance and value should be included; there should be adequate weight given to commercially valuable species, as to socially and culturally significant species. Extremely lucrative fisheries or those where there are many stakeholders/partners may provide insight into the issue of political economy and policy coherence. Cases that have employed market-based mechanisms or economic incentives towards fisheries rebuilding must also be included; for example, how have such incentive been applied to effectively manage bycatch and/or discards?

Box 3.1. Case study template

In conducting the case studies, the OECD draws upon the expertise of Member-country authorities and external experts. To focus the data gathering and analytical efforts of a varied group of researchers, a detailed and comprehensive template was prepared to guide the collection of information and review of the case studies. This template ensured that the information collected was focused on the economic, social and institutional factors associated with each rebuilding plan.

Each case study includes a short description of why it was chosen, the key characteristics of the fishery, and a short statement about the institutional context, as well as a description of major stakeholders. The case study includes an overview of the design, structure and implementation of the fisheries rebuilding plan. The key elements of the template to gather information and data on the case studies of rebuilding fisheries were follows:

- Background: This refers to the key facts relevant to the fishery that underscore the rebuilding plan, and basic contextual information on the institutional framework. This section also provides details on the rebuilding plan and approach.

- Economic and social aspects: Included here are economic instruments used to support the rebuilding process and other relevant economic information. A description of the key stakeholders, how they were involved in rebuilding, as well as distributional issues and any compensation packages or programmes to manage a transition would be described here.

- Implementation issues and lessons learned: This section is meant to obtain information on the political economy of the rebuilding process, including identifying obstacles and tradeoffs as well as how they were overcome. Best practices may eventually be identified from this section.

- Annex: Basic indicators of the fishery subject to rebuilding are included, as well as a profile of the fishing industry. This may yield valuable information on the progress of these indicators throughout the rebuilding period and the evolution of the industry(ies) involved in the fishery in response to the rebuilding measures.

- *Policy coherence*: Cases that highlight mutually reinforcing policy actions across government departments towards achieving agreed objectives, as well as cases that demonstrate the challenges that arise when there is a lack of coherence, are considered.

- *Political economy*: Examples that illustrate how political economy issues were addressed in the development of rebuilding plans include how distributional aspects were tackled as well as the role of stakeholders.

- *Successful vs. unsuccessful:* The choice of case studies are not be geared towards selecting so called "success stories". Rather, this exercise is about identifying examples of good practices and, to the extent possible, identifying what did not work and why. In addition, recent rebuilding plans should not be excluded simply on the basis that results are not yet visible; they may in fact provide useful information in terms of the design and implementation process as they may result from lessons learned from previous rebuilding plans and represent a course correction.

Literature review

Several recent studies examining the challenge of rebuilding depleted fish stocks utilised a case study approach with a view to drawing out best practices or developing guidelines for effective fisheries rebuilding plans. These studies are described here in order to provide a summary of previous research on this topic.

An overview of recent global experience with recovery plans for depleted marine resources and suggested guidelines for recovery planning
by Caddy and Agnew (2004)

One of the first major overviews of stock rebuilding programmes was undertaken by Caddy and Agnew (2004). The study is based on an invited plenary lecture to the 2003 ICES Annual Conference and reviews eight case studies of successful and unsuccessful stock rebuilding programmes from the United States, Canada, New Zealand and the European Union. It develops a number of insights from the reviews and proposes tentative guidelines for best practice in fishery recovery plans.

This study approached the review of fisheries rebuilding cases in two ways: through a high level review of all plans targeted at rebuilding fisheries, using publicly available information whether in the form of a formal plan or a series of measures aimed at rebuilding. Second, a detailed assessment was made of eight cases from Canada, the United States, New Zealand and the European Union for Pacific Halibut, Gulf of Mexico King Mackerel, Striped Bass, Summer flounder, Pacific Ocean Perch, Canada [Atlantic] cod, Canadian haddock and Cod in the North East Atlantic.

The appendix lists 67 points of consideration for best practices in fishery recovery plans in six categories: actions prior to the recovery process; issues to be considered by the recovery team; recovery objectives; recovery management; and post recovery. These considerations focus primarily on biological advice, research and assessments, management processes, and monitoring and evaluation. There is also recognition of the political economy issues (e.g. political pressures post recovery) and the importance of consensus and negotiation with stakeholders. However, economic considerations or market-based approaches are not examined in detail.

An evaluation of rebuilding plans for US fisheries, Lenfest Ocean Programme
by Swasey and Rosenberg (2006)

Swasey and Rosenberg (2006) undertook a major evaluation of the rebuilding plans for depleted stocks in the US and the results are summarised in Rosenberg *et al.* (2006). The study provides a detailed scientific review of the rebuilding plans and management for 67 fish stocks. It was found that, as of 2005, overfishing (where the fishing mortality rate exceeds the level that should support MSY) continued in 45% of the stocks under rebuilding plans and around 72% of stocks remained overfished. Three stocks had been rebuilt, but fish stock abundance appeared to be increasing in 48% of the stocks under rebuilding plans. The study methodology was based on publicly available data, and was assisted in its execution by the availability of precautionary reference points as required under the US fisheries legislation.

Review of institutional arrangements and evaluation of factors associated with successful stock recovery programmes
by UNCOVER

UNCOVER[1] is a major project funded by the European Commission that seeks to develop insights into strategies for stock rebuilding in a number of fisheries. The objective is to identify changes experienced during the decline of fish stocks, to enhance the scientific understanding of the mechanisms for fish stock recovery, and to formulate recommendations for fisheries managers on how to best implement stock recovery plans.

Four case study areas are analysed: Barents and Norwegian Seas (covering NE-Arctic cod, Norwegian spring spawning herring and capelin); North Seas (cod, plaice and autumn spawning herring); Baltic Sea (sprat and Eastern Baltic cod); and Bay of Biscay (Northern hake and anchovy). The overall work plan for the project is focused on modelling alternative strategies for stock recovery in the case study areas. It includes an economic component that focuses on bio-economic modelling of selected stocks, the development of four community socio-economic profiles for Spain, France, the Netherlands and Scotland, and a social impact assessment of one of the recovery strategies on a pilot scale for Denmark.

As part of the UNCOVER project, the Marine Resources Assessment Group (MRAG) undertook a review of institutional arrangements and the key factors associated with successful recovery plans (Wakeford *et al.*, 2007). The study reviews 33 case studies from the United States, Australia, New Zealand and Europe and used 13 performance criteria to evaluate the relative importance of institutional, economic, social and environmental factors in stock rebuilding plans. Amongst the key findings from the study, the authors found that recovery is effective under the following conditions.

–   Catches are significantly reduced over a short period of time, creating a positive shock to the stock.

–   The recovery plan is part of a legal mandate which is automatically triggered on reaching pre-defined limit reference points.

–   The economic efficiency of the fleet is evaluated and monitored throughout the rebuilding process.

–   Effort reductions are created using input controls in addition to TAC reductions, rather than through output controls.

Recovering Canadian Atlantic cod stocks: The shape of things to come
by Rice et al. (2003)

Rice *et al* (2003) analysed the collapse of North West Atlantic groundfish stocks in the 1990s and note the following key observations and lessons learned.

–   The potential for recovery is variable by stock so management approaches should be tailored accordingly, including the assessment of the economic impact. Some Atlantic cod stocks reacted favourably to moratoria and sustained commercial fisheries for a time, but then declined. Others have consistently remained at low biomass levels.

–   If the underlying issues that led to overfishing are not addressed, such as the permanent removal of excess capacity, the risk of overfishing will reoccur should the stock recover.

    –   Deferral of rapid and decisive management action to reduce harvest because of uncertainty about stock status and concerns about the impacts of the reductions on the fisheries contributed to the severity of the collapse and ultimate severity of the measures needed to commence recovery.

These studies illustrate that the previous studies on rebuilding focussed on the recovery of stocks. The OECD study aims to build on this to examine rebuilding fisheries, which includes a healthy stock, ecosystem and industry through the early inclusion of economic analysis and market based measures.

Main observations

A total of 23 case studies have been done for this project.[2] These case studies reflect fisheries rebuilding plans and/or activities in OECD countries, developing countries, and those led by Regional Fisheries Management Organisations (RFMOs).

Table 3.1. Fisheries rebuilding case studies

Species	Country
Snow crab (Chionoecetes opilio)	Japan
Sailfin Sandfish	Japan
Chum Salmon	Japan
Sailfin Sandfish	Korea
Swimming Crab	Korea
Yellow Croaker	Korea
Cephalopods (octopus)	Mauritania
Hake	Namibia
Bluefin Tuna	CCBST
Greenland Halibut	NAFO
Herring and sprat	Estonia
Cod	Iceland
Capelin	Iceland
Herring	Iceland
Abalone	Mexico
Red grouper	Mexico
Pink shrimp	Mexico
Queen conch	Mexico
Hoki	New Zealand
Scallops	France
Groundfish	Canada
Cod	Sweden
Cod	Denmark

Integrating economics early

The Korean case studies note that limited information on the economic impact on the rebuilding plan can be a factor in the resistance by stakeholders, which in turn impede the effective implementation of the plan even if specific rebuilding measures are put into place. In Korea, education on the basis of the plan coupled with consultation are emphasised as a means of overcoming this challenge, although it is also acknowledged that education and communications activities alone are not sufficient to tackle the challenge if the right incentives are not in place.

A key lesson from the Namibian hake case study is that social aspects require consideration during the design of rebuilding plans and that establishing the social success of a plan should be an objective and not only as a spill-over of economic success. In this case, employment levels and the redistribution of profits were taken into account.

The sailfin sandfish case in Japan illustrates that rebuilding a stock may not always immediately lead to a strong industry. A three-year closure of the fishery was imposed, resulting in significant recovery of the population. However, an unintended consequence was that the price of the fish decreased and economic returns were marginal.

The case studies from New Zealand and Iceland illustrate that rights-based management (RBM) systems may be effectively used to rebuild fisheries. Although they have mostly been discussed in relation to generating rent and reducing fleet capacity, they are also effective in rebuilding fisheries on the brink of collapse. RBM systems are driven by economic incentives but seem to be suitable to achieve other goals such as rebuilding fish stocks.

The rebuilding of the scallop fishery in St Brieuc, France was made easier due to the fact that fishermen's concern for their own profits helped to create a consensus for a rebuilding plan.

In the Canadian Pacific Groundfish programme, positive experience from earlier ITQ programmes helped in paving the way for the introduction of ITQs for rebuilding.

The Swedish case study on Baltic Sea cod demonstrates that increased landings are not sufficient to increase benefits in a fishery. Overcapacity must also be addressed. The same is true for the Danish Baltic Cod case where increased rents are mainly due to reduction of overcapacity rather than increased catches. This case study also highlights the importance of taking into consideration additional complexities when dealing with multi-species fisheries. If a bigger stock is not accompanied by increased flexibility in the fishery, the economic gains from increasing the biomass of a single stock may be small.

The importance of stakeholder involvement

Close collaboration with stakeholders in designing rebuilding plans and instituting measures is emphasised in several case studies. The Korean approach includes regular review and evaluation of the plans in consultation with stakeholders so that appropriate course corrections can be made as needed. The Namibian hake case study notes the importance of political will and support from national authorities as a key driver for success in the implementation a rebuilding plan.

In Japan, the initial reaction of fishers to proposed rebuilding measures were negative. For example, fishers opposed certain measures that would be instituted for the first time (e.g. concrete blocks) in the snow crab fishery primarily because their effects were unknown. To mitigate these concerns, an incremental approach was pursued where the biological effects of instituting one marine reserve were monitored and regularly communicated to fishers. Once fishers realised that stocks increased (and hence catch), the opposition to this measure declined.

The Korean yellow croaker case study illustrates the complexity of multi-species fisheries, wherein actions involving the directed fishery alone may not be sufficient for rebuilding. Given the potential conflicts among different segments of the fisheries, co-ordinating various interests generally presents a challenge for fisheries managers.

Effective communication between researchers and fishers is also crucial to rebuilding, as illustrated by the actions of the fishery research institute of Akita Prefecture in the sailfin sandfish case. Initial population models for the sailfin sandfish projected that catch would triple after a three-year closure of the fishery, while the catch actually increased more than originally projected. Key scientific information was shared by relevant stakeholders, and the process established trust between local fishermen and the research institute which assisted in instituting a fishery closure.

In Canada, a new structure was set up to efficiently engage stakeholders in the design of the rebuilding plan and obtain their buy-in. The Fisheries and Oceans Canada (DFO) provided a broad set of guidelines and requirements for the outcomes, while it was left to the stakeholders to decide on the specific nature of the programme to reach those objectives. In that way stakeholders were given responsibility while at the same time being empowered. Getting stakeholders to participate in the design of the rebuilding plan was very successful, especially by taking into account the number and different characteristics of stakeholders in these specific fisheries. It probably also prompted stakeholders to participate in the design of the rebuilding plan in that if they did not participate there was the indirect threat that a moratorium would be imposed with serious consequences for all stakeholders involved.

The use of flanking measures to support rebuilding objectives can ease the transition. In the sailfin sandfish case study, it was noted that as part of the agreement on the three-year fishery closure, the prefectural government provided incentives to those fishers who complied with the self-imposed (voluntary) regulation. This included decommission subsidies for inactive vessels and gear, low interest rate loans, and additional scientific research. Another *de facto* incentive provided by the government was the continuation of the limited entry system for the fishery. As there would be no new entrants to the fishery after the rebuilding, the expected benefits of the fishery closure (even though the exact amount of the future benefit was largely unknown) would be received by the same fishers who bore the costs of the closure.

The Icelandic experience shows that having the possibility of distributing quotas to hard hit regions or sub-sectors of the fishery may contribute to the sustainability of the fisheries management system. Widespread disagreement on distributional issues may undermine rebuilding plans and can have an effect on the probability of success and survival of such plans. Flanking measures may be necessary in RBM rebuilding plans to guarantee support. The Danish case study indicates how limitations on the transferability of rights between bigger and smaller vessels may impede apparent negative distributional effects.

Incremental approaches to rebuilding fisheries

The Korean cases demonstrate how an incremental approach to rebuilding can be undertaken in situations where full data for decision making is unavailable. Rather than pursuing more concrete scientific evidence, rebuilding plans were established. These are subject to regular reviews and modified based on monitoring and evaluation exercises. This demonstrates that immediate and early rebuilding efforts are an important feature of Korean rebuilding plans, and this incremental strategy could be one way to rebuild fisheries in data-limited circumstances while following the precautionary approach. The Southern Bluefin Tuna case study notes that "the longer the delay and inaction, the higher the probability that the rebuilding will be unsuccessful, the greater the cost, and the greater the possibility of a stock collapse".

The necessity for a broad range of management measures

The Japanese and Korean case studies illustrate that addressing catch levels is not the only solution to rebuild some stocks. Habitat improvements and stock enhancement may need to be implemented as part of a broader range of measures in a rebuilding plan, particularly if these in areas that pose the most threat to the rebuilding of the species. This also holds true in other countries, and is particularly relevant for species such as salmon and eels (e.g. European eel). By the same token, the Japanese salmon case also indicates that that stock enhancement programmes should be implemented together with appropriate fishing regulations for a comprehensive approach to stock rebuilding. As noted in other studies, management measures must be accompanied by favourable environmental factors in order to be successful and that a holistic approach that addresses various threats to the species should be examined.

The Estonian case study illustrates that great improvements can be taken towards rebuilding a troubled fishery by changing the institutional structure of the industry. The challenges facing fisheries in need of rebuilding are not solely related to small stocks or low recruitment, but also with processing, transport, marketing, and the horizontal and vertical integrations in the value chain.

Monitoring and enforcement: key elements of a rebuilding plan

Both the Korean and Namibian case studies emphasised enforcement of rebuilding measures. In the case of the Korean swimming crab, the management committee placed particular emphasis on monitoring crab markets and investigating transactions involving illegal harvest of crabs (e.g. undersized crabs). Local governments, fisheries co-operatives and other representatives of fishers jointly monitor major fish markets on a regular basis. Nevertheless, monitoring and enforcement is still a challenge due to limited resources.

In the case of Namibia, monitoring, control and surveillance (MCS) measures were recognised as a key to the success of a rebuilding plan, also supported with a legislative framework (e.g. to include specific fines etc) and appropriate resources. In the Namibian hake fishery, Monitoring and control was greatly facilitated by the fact that there are only two landing sites and the fleet is industrialised.

While some fisheries are data rich and have advanced monitoring and surveillance systems, other do not. The Mexican and Turkish case studies underline the importance of using the knowledge and resources of the fishing communities to alleviate such problems. The Japanese experience with co-management and the use of TURFs demonstrate how fishers themselves can help in the monitoring and surveillance activity necessary for rebuilding.

Good quality of data and efficient control and surveillance was also a key element in the successful rebuilding of the scallop fishery in St Brieuc, France.

In Canada, the setting up of an efficient monitoring system was an integral part of the rebuilding plan and a key to its success. Although costly, it is unlikely that it would have been possible to obtain better data and keep the fishing mortality within acceptable limits without setting up such a system.

Trans-boundary stocks require joint and co-ordinated efforts

The Greenland Halibut and Southern Bluefin Tuna case studies emphasise the need for actions that are not only agreed to by all relevant parties, but are also adhered to by fishers from all countries. In the case of swimming crabs which migrate across both Korean and Chinese waters, it has been noted that efficient management in only one nation is not sufficient to rebuild the stock. In addition, co-operation in terms of developing stock assessments and coherence across rebuilding measures is also required.

The Danish Baltic Cod fishery case further demonstrates the complexities of rebuilding fisheries which are not only harvested by many countries, but are also multi-species fisheries. This raises problems of how to account for unavoidable by-catch. It also shows how national policies can differ when decisions on TACs and technical measures are decided at a supra-national level, while decisions on management systems are made at the national level.

Notes

1. UNCOVER is the acronym for the full name of the project, Understanding the Mechanisms for Stock Recovery. The project is a consortium of 17 fisheries research organisations across Europe and is scheduled to be finished in February 2010.

2. See OECD Food, *Agriculture and Fisheries Working Papers* series (*www.oecd.org/fisheries*).

References

Caddy, J.F. and D.J. Agnew (2004), "An Overview of Recent Global Experience with Recovery Plans for Depleted Marine Resources and Suggested Guidelines for Recovery Planning", *Review of Fish and Fisheries*, Vol. 14, pp. 43-112.

Costello C., B.P. Kinlan, S.E. Lester, and S.D. Gaines. (2012), "The Economic Value of Rebuilding Fisheries", *OECD Food, Agriculture and Fisheries Working Paper N°55*, OECD, Paris.

Feagin, Joe R., Anthony M. Orum, and Gideon Sjoberg (1991), *A Case for the Case Study*, University of North Carolina Press.

Rice, J.C., P.A. Shelton, D. Rivard, G.A. Choninard and A. Fréchet (2003), *Recovering Canadian Atlantic cod stocks: The shape of things to come?* ICES CM U:06.

Swasey, J.H. and A.A. Rosenberg (2006), *An Evaluation of Rebuilding Plans for US Fisheries*, Lenfest Ocean Program, Washington DC, April, available at *www.lenfestocean.org\publications.*

Wakeford, R.C., D.J. Agnew and C.C. Mees (2007), *Review of Institutional Arrangements and Evaluation of Factors Associated with Successful Stock Recovery Programs*, CEC 6th Framework Programme No. 022717 UNCOVER, MRAG Report, March.

Yin, Robert K. (ed.) (2004), *The Case Study Anthology*, Thousand Oaks, CA: Sage Publication Inc..

Chapter 4.

National and international approaches to rebuilding fisheries: A synthesis

This analysis is based on an inventory of national and international policies that guide rebuilding programmes. The inventory provides a comprehensive overview of rebuilding policies and helps in sharing information on different policy frameworks and approaches. The analysis highlights the challenges in managing international fisheries and provides valuable insights for policy makers. The role of stakeholders is of central importance. Coherence across legislative provisions and policy tools is essential, especially as it builds trust among stakeholders. Incoherence across policies applied in the fisheries sector undermines policy objectives and is counterproductive. In the same way, it is important to plan early in a post-rebuilding management strategy as this will provide certainty for stakeholders. Furthermore, transition mechanisms, including flanking measures, may be needed to obtain and maintain support for necessary reforms.

Strategies and approaches to effectively rebuild fisheries that meet biological objectives and that take into account social and economic considerations figure prominently in the present policy debate. This is against the backdrop of stagnating wild fish harvests at a global level coupled with increased demand for food, as well as issues such as the need for food security and mitigation of the effects of climate change. Rebuilding plans, if well designed, can work towards the goal of sustainable fisheries that are characterised by a resilient ecosystem coupled with lasting economic opportunities.

Countries have committed to international principles and targets, and have developed national approaches to address rebuilding fisheries. In particular, at the World Summit on Sustainable Development in 2002, governments committed to the goal of rebuilding fish stocks to levels that can produce the maximum sustainable yield by 2015. Recognising that rebuilding international fish stocks must occur within a cooperative governance framework, the UN Fish Stocks Agreement (UNFSA, December 2001) has enabled Regional Fisheries Management Organisations (RFMOs) to address the overfishing of straddling and highly migratory stocks fished primarily on the high seas.

While governments have committed to "maintain or restore stocks to levels that can produce the maximum sustainable yield with the aim of achieving these goals for depleted stocks on an urgent basis and where possible not later than 2015" as part of the Plan of Implementation of the World Summit on Sustainable Development, relatively little is known about how they implement strategies to meet this goal.

This chapter provides a review of the experiences of OECD countries in the design, implementation and outcomes of fisheries rebuilding programmes. It also provides an indication of the information being gathered for the inventory of national rebuilding programmes on the legislative and policy basis for rebuilding fisheries, the biological and socio-economic information collected and analysed for decision making, and the management standards and requirements specific to each national approach. This information is based on research conducted by OECD, and data provided by OECD countries.

Scope of rebuilding fisheries policy approaches

This chapter provides a summary of national and regional approaches, policies and guidelines that are relevant to fisheries rebuilding plans, with a focus on institutional and economic aspects. Approaches are being investigated at the following levels.

- *National*: Programs and policies of individual countries in order to gain insights on the institutional structures and legislative frameworks guiding rebuilding efforts, as well the type of economic information collected and how it is considered in rebuilding plans.

- *Regional*: This follows the framework for fisheries rebuilding under the Common Fisheries Policy for EU member states. In addition, a review of RFMO methods to fisheries rebuilding may provide constructive information on approaches and lessons learned in an international context in which responsibility is shared.

The objective is to develop a comprehensive overview of the policies guiding fisheries rebuilding programmes involving countries at the national and regional levels, with a particular focus on institutional and economic factors. The OECD Committee for Fisheries considers it useful to undertake an inventory of fisheries rebuilding related activities for the following reasons.

- *Sharing information:* As a result of the declining number of commercial capture fisheries, coupled with the increasing demand for fish and seafood, developing effective rebuilding plans continues to be a significant challenge facing many countries. By combining and sharing knowledge and experience through the development of an inventory of approaches and best practices, countries have a better chance to develop synergies and apply lessons learned to address the challenge of rebuilding fisheries successfully.

- *Focus on economics:* Highlighting economic and institutional aspects of fisheries rebuilding provides value added to the body of knowledge on rebuilding fisheries which has largely had a biological focus, although some also examine management structures and economic mechanisms. The focus of this inventory will allow for a greater understanding of political economy issues involved in rebuilding fisheries at the national and regional levels, and will include the governance system, the role of regulation, and the interplay between stakeholders, fisheries managers and others. This will allow for a better understanding of how countries integrate a broad set of information, from ecological to economic, into the decision-making processes.

- *International co-operation:* Management of straddling stocks and fisheries on the high seas is complex and often requires the involvement of several countries. An understanding of how co-operative structures work to manage depleted fisheries will provide information on the issues, challenges and possible solutions in an international context.

Analytical framework

A template was designed to collect information from member countries on the legislative and policy basis for rebuilding fisheries, the biological and socio-economic information collected and analysed for decision making, as well as the management standards and requirements specific to each national approach.

It is recognised that compiling information with respect to the life cycle of the fisheries rebuilding process is a huge undertaking that requires input from diverse national experts in the fields of biology, fisheries management, and economics. In addition, it may be difficult to obtain some information requested as countries may not always have distinct policies, processes or procedures for each section of the template (e.g. there will be differences in approaches for regional entities as opposed to national systems, given the division of responsibilities between an organisation and its members). As such, an iterative approach was pursued whereby countries were consulted when additional information or clarifications were needed.

Box 4.2. Economics of rebuilding: Template on national policies and approaches

Context: This section provides an overview of the status of fish stocks within a country or region.

Legislative and policy framework: This section outlines the legislative basis for rebuilding, as well as supporting policies and guidelines, and could be informative for other countries in the development or review of their policies. Information regarding the definitions of key terms associated with rebuilding were requested.

Scientific framework: This section describes the basic scientific foundation for fisheries rebuilding plans.

Rebuilding plans: This section describes the framework and structure for rebuilding plans.

Economic aspects: This section is intended to capture information on how economic considerations are taken into account in the decision making process, as well as provide information of the use of economic tools and approaches (e.g. market-based mechanisms used to support rebuilding).

Social aspects: This section describes consultation and collaboration with stakeholders, and the use of mechanisms to promote rationalisation of fishing fleets.

International context

Governments worldwide have sought to address the issue of depleted and overfished stocks through various international agreements to enable a prosperous and thriving fishing sector. To that end, political commitments have been made through a series of hard (binding) and soft laws (non binding).

With the adoption of the United Nations Convention on the Law of the Sea (UNCLOS) in 1982, coastal states were provided with jurisdiction over a 200-nautical mile exclusive economic zone (EEZ), within which they are required to protect aquatic resources against overfishing. This was considered to be an important step to enable countries to protect and conserve stocks. Building on this framework, there have been numerous complementary agreements that seek to establish standards for fish conservation and management on a global scale. Key international accords include the following.

- The 1995 United Nations Fish Stocks Agreement focuses on the conservation of straddling and highly migratory fish stocks by expanding on UNCLOS, including the responsibility to apply the precautionary approach by setting limit reference points for maximum sustainable yield. This provides a foundation for regional fisheries management organisations (RFMOs) with respect to rebuilding stocks.

- The 1995 FAO Code of Conduct for Responsible Fisheries is a non-binding instrument that has been accepted by all 188 members of the FAO. This code states that overfishing should be prevented, along with excess fishing capacity, and that sustainable management measures be promoted. This code demonstrates the commitment by all member nations to the importance of rebuilding depleted fisheries.

- The pressure to address overfishing was discussed at the World Summit on Sustainable Development in 2002, where governments committed to an ambitious goal of rebuilding fish stocks. Specifically, the Johannesburg Plan of Implementation requires countries to "maintain or restore stocks to levels that can produce the maximum sustainable yield with the aim of achieving these goals for depleted stocks on an urgent basis and where possible not later than 2015".

Overview of national approaches in OECD countries

This section presents an overview of national approaches to fisheries rebuilding in a selected number of countries. It is clear from this overview that the approaches to rebuilding fisheries differ between countries, sometime significantly. This highlights the fact that there is not a "one size fits all" approach to the issue of designing and implementing cost-effective and efficient fisheries rebuilding plans, and that considerable attention needs to be paid to the range of ecological, economic, social and institutional characteristics underlying each country's circumstances. It also underscores the value of undertaking this exercise as a wealth of information is shared.

Chapter 2 notes that defining terminology in an international context is important. For example, some countries refer to "recovery" (European Union) or "restoration" of fish stocks. However, in other countries, there may be legal connotations regarding such terminology; for example, "species recovery plans" is a specific term referring to plans and actions directed to threatened or endangered species (United States, Canada). The approaches to fisheries rebuilding also vary from specific plans directed at particular stocks to others that integrate rebuilding objectives as one component within fish management plans.

Rebuilding fisheries: Taking stock

Charting progress of national rebuilding plans and stock status improves transparency and can help articulate the benefits of the plans

Taking stock of national fisheries is conducted through a regular reporting exercise in some countries, and often includes an overall assessment on the number of stocks that are depleted, overfished or subject to overfishing, as well as accounting for the status of rebuilding plans. The level and detail of information on the status of fish stocks and rebuilding plans varies across countries. The United States is legally obliged to produce an annual report on the status of its fish stocks which provides details on several aspects. New Zealand and Australia produce regular, detailed assessments on their fisheries that are readily available in a single publication. Canada also has a comprehensive system of taking stock of national fisheries.

The National Oceanic and Atmospheric (NOAA) Fisheries in the United States is required to produce an annual report to Congress reporting on the status of US fish stocks. As of the last quarter of 2009, 52 rebuilding plans were active in the United States. Currently, 57 stocks or stock complexes have overfished thresholds not defined or applicable, or are unknown with respect to their overfished status while NOAA Fisheries has adequate information to determine the status of 173 fish stocks and, of these 129 stocks or stock complexes are not overfished (four of these stocks are approaching an overfished condition) while 44 stocks or stock complexes are overfished.

In New Zealand, stock status have been summarised annually since 2006. As of September 2009, sufficient information is available to describe stock status relative to

MSY-compatible targets for 117 of the 628 fish stocks in the country's quota management system. This represents a net increase of 16 stocks (15.8%) over the 101 stocks of known status a year earlier. Stocks of known status accounted in 2011 for 72% of the total landings by weight and value – up from 66% a year earlier – and represent most of the main commercial species. Of the 117 stocks or sub-stocks with known status relative to target reference points, 79 (68%) were determined to be near or above target levels based on a recent assessment or evaluation, while the remaining 38 (32%) stocks are known to be below their respective targets.

In 2008, 98 fish stocks in Australian Government-managed fisheries were assessed in terms of biological status (overfished status and overfishing status); the number of stocks assessed has increased steadily from 31 in 1992 to 98 in 2008 (97 in 2006 and 96 in 2007). The number of stocks classified as not overfished or subject to overfishing increased to 27 in 2006, 28 in 2007 and 39 in 2008, following a five-year period in which they remained stable at around 18 to 20. In 2008, 18 stocks were classified as either overfished and/or subject to overfishing, up from 16 in 2007, but down from 19 in 2006. From 1996 to 2005, the number of stocks classified as overfished and/or subject to overfishing increased steadily from three, to a peak of 24 in 2005. The status of stocks in Australian Government-managed fisheries is reported annually in *Fisheries Status Reports*.

In Canada, the *Fishery Checklist*, developed in 2007, is used as a self-diagnostic tool to monitor improvements in the management of a fishery, as well as gathering information on major stocks and their fisheries. The *Checklist* has 106 questions that cover, for example, scientific issues such as stock statues and the presence of reference points as well as fisheries management and enforcement issues. Although not designed as a tool for public reporting, the results and specific indicators from the *Fishery Checklist* have been used to report and gauge progress on various issues.

In its communication *Consultation on Fishing Opportunities for 2010* (May, 2009), the European Commission indicated that the status of some 59% of stocks is unknown. Of those for which the state of stocks is known, 69% are at high risk of depletion and only some 31% of stocks are known to be fished sustainably. Since 2002, management plans have been developed for many stocks: 41% of pelagic stocks (41% of catches) and 29% of demersal stocks (44% of catches) are now under multi-annual plans. Work will continue on bringing more stocks under such plans, including the pelagic stocks in the Baltic Sea and a few Mediterranean fisheries. Specific plans will be proposed in 2009 for northern hake, western horse mackerel, Bay of Biscay anchovy and Baltic salmon. Ten plans are implemented and another six were in development for 2009/10.

Over the last twenty years, Japan has increased its monitoring of its primary fish stocks. The Japanese public research institute classifies assessed stocks into three categories (high, middle, low) in terms of the relative abundance. The 2004 assessment indicates that the resource levels of 12 stocks, including saury, common squid and sea-bream, are classified as high, 49 fish stocks such as common mackerel, sardine, Alaska pollock, and snow crabs are low and 30 stocks, including Jack mackerel and sand fish, are classified as middle. As of February 2008, 51 plans for specific fish species and 20 comprehensive plans covering geographic areas and fishing types were developed or were under development.

Table 4.1 Summary of stock status and rebuilding plans

Country	Stocks assessed	Rebuilding plans	Status of known stocks
Australia	98		18 Overfished/overfishing
European Union		10	69% At high risk of depletion
New Zealand	117		38 Below target
United States	173	52	44 Overfished/overfishing
Japan	92	71	49 Classified as low

Source: Country submissions

Legislative and policy framework

Legislative and regulatory support provides a strong foundation for rebuilding while policies and guidelines provide direction

A country's institutional and legislative arrangement plays a role in fisheries rebuilding. Understanding the legislative conditions is important to set the stage for rebuilding plans.

National legislation mandates the rebuilding of fisheries in the United States, along with specific timelines and limited flexibility. Caddy and Agnew (2004) and Wakefield *et al. (*2007) state that rebuilding success is more likely in jurisdictions that have explicit legislation. However, other studies indicate that some flexibility is required in order to integrate economic factors in the objective setting exercise, establishing time frames for rebuilding and developing strategies to mitigate the impact of rebuilding measures (Larkin *et al.*, 2007).

Larkin *et al* (2007) contrasted the approach to rebuilding in the United States with the more flexible one employed in New Zealand. Their research indicates that the ability to adjust a rebuilding timeframe according to a broader set of goals, including socio-economic objectives, could increase the net present value of commercial harvests. A more flexible approach may allow for rebuilding plans that meet biological targets in a socio-economical optimal way, while continuing to actively engage the users of the resource in the decision-making process. On the other hand, supporters of the existing US legislative rebuilding provisions state that the ten-year timeframe is feasible in practically all situations, that adequate exceptions are allowed, and that strong and early actions to rebuild overfished stocks make greater economic sense over the longer term. In summary, legislated rebuilding requirements continue to generate controversy in the United States.

Supporting policies can provide clear and transparent guidance regarding the design and implementation of rebuilding plans. The benefit of supporting policies is that they are more flexible than legislation, can be more easily amended or updated over time, and can respond to changing circumstances or emerging issues. Several of the respondent countries summarised such policies and guidelines (*www.oecd.org/fisheries)*. In several countries, there exists complementary legislation and policy guidance that accompany the major legislation.

Legal mandate

In the *United States*, stock rebuilding was first mandated under the 1996 Sustainable Fisheries Act which amended the Magnuson-Stevens Fishery and Conservation Act (MSA), and by the more recent 2007 amendments to the same law. These amendments provide a legal mandate for NOAA Fisheries to end overfishing and rebuild overfished fish stocks. Rebuilding plans are normally undertaken as an amendment to an existing Fisheries Management Plan, which are developed by the regional fisheries management councils and implemented by NOAA Fisheries. There are also provisions to use Limited Access Privilege Programmes as a market measure to help rebuild overfished stocks and for the use of annual catch limits to ensure that overfishing does not occur

The MSA mandates that once a stock has been determined to be overfished, it must be rebuilt in as short a period as possible, not to exceed ten years, with certain exceptions. The mandatory ten-year rebuilding time-frame has prompted criticisms that the law is excessively rigid and unrealistic. As a consequence, some members of Congress have recently introduced *The Flexibility in Rebuilding American Fisheries Act of 2009* citing unfairness and excessive consequences on fishing communities resulting from the limitations placed on fisheries managers to meet legislative timelines.

In the United States, legal authority to recover listed species and selected marine mammals is also provided to NOAA Fisheries under the Marine Mammal Protection Act and the Endangered Species Act. These are complementary legislations that are primarily directed at protecting endangered species at risk of extinction.

The challenge to rebuild depleted stocks has been taken up at the regional and national level to varying degrees and through various mechanisms. In the *European Union*, Council Regulation (EC) No. 2371/2002 of 20 December 2002 on the conservation and sustainable exploitation of fisheries resources under the Common Fisheries Policy enables the Commission to take steps to rebuild fish stocks or establish emergency measures. For the Mediterranean area, supportive legislation is provided by Council Regulation (EC) 1967/2006 that, inter alia, creates a framework and obligations to set up multi-annual management plans in line with the basic 2371/2002 regulation. In 2008, the Commission launched a review of the Common Fisheries Policy (CFP) which was based on an analysis of the achievements and shortcomings of the current policy, and looked at experiences from other fisheries management systems to identify potential avenues for future action. A new CFP is to enter into force in 2013 (*ec.europa.eu/fisheries/reform/index_en.htm*).

In *New Zealand*, Section 13 of the Fisheries Act requires the Minister of Fisheries to set a total allowable catch that "maintains the stock at or above a level that can produce maximum sustainable yield, having regard to the interdependence of stocks". For stocks that are below the level that can produce the maximum sustainable yield, the Minister must set a total allowable catch that enables the level of the stock to be altered "in a way and at a rate that will result in the stock being restored to or above a level that can produce the maximum sustainable yield, having regard to the interdependence of stocks; and within a period appropriate to the stock, having regard to the biological characteristics of the stock and any environmental conditions affecting the stock."

Section 13 of the New Zealand Fisheries Act was amended in February 2008 to incorporate situations where it is not possible to explicitly estimate current biomass or B_{MSY}. In such cases, the Minister must set a total allowable catch that is "not inconsistent with objective of maintaining the stock at or above, or moving the stock towards or above, a level that can produce the maximum sustainable yield". These sections of the Fisheries legislation are further elaborated in the Harvest Strategy Standard.

In *Norway*, the primary legislation for the management of the fisheries is the Act relating to the Regulation of the Participation in Fisheries and the Act relating to the management of wild living marine resources. Fish stock rebuilding takes primarily place under the Act relating to the Management of wild living marine resources. However, in special cases with a threatened and endangered marine species, this species can be prioritised according to the Nature Diversity Act. This Act sets out requirements to protect and implement recovery strategies for the species.

There is no law/regulation comprising specific provisions on fisheries rebuilding in *Turkey* under the existing national fisheries management regime. However, the national legislation includes provisions on the conservation of living marine resources. The Fisheries Law-1380, the primary law laying down management and implementation rules for fisheries and aquaculture empowers the Ministry of Agriculture and Rural Affairs to collaborate with private agencies, universities, research institutions and international organisations to increase productivity and the conservation of natural stocks and to protect them from biological and non-biological threats. The Implementing Regulation on Fisheries 1995 is the fundamental regulatory instrument for marine and inland fisheries. The regulation covers rules, *inter alia*, on fishing gears, prohibitions, limitations, inspection and control.

The *Korean* Government has set fish stock rebuilding as its main fisheries policy objective. In this context, the government established and announced the Fisheries Resources Management Act (FRMA) in April 2009 with a view to establishing and implementing fisheries resource recovery plans. The objectives of the FMRA are to strengthen fisheries research and assessment capabilities, establish and implement fish stock rebuilding plans, and to continue to implement key fisheries management tools such as stock enhancement. The Act incorporates the relevant sections from the conventional *Fisheries Act* regarding protection and management of resources, with features related to stocking from the *Promoting Nurturing Fisheries Act*.

In *Australia*, Commonwealth fisheries are managed under the Fisheries Administration Act 1991 (FA Act) and the Fisheries Management Act 1991. Under these acts, the fisheries minister and Australian Fisheries Management Authority (AFMA) must pursue objectives relating to ecological sustainability of target and non-target species, maximising net economic returns, ensuring that the living resources of the Australian Fishing Zone are not endangered by over-exploitation, achieving optimum utilisation of the living resources of the Australian Fishing Zone, and complying with obligations under international agreements. In December 2005, a Ministerial Direction was issued to AFMA under section 91 of the Fisheries Administration Act 1991 to recover overfished stocks, develop a world's best practice harvest strategy policy for Commonwealth fisheries, and investigate the use of individual transferable quotas in the management of all fish stocks in Commonwealth fisheries. The Environment Protection and Biodiversity Conservation Act 1999 (EPBC

Act) is Australia's primary environmental legislation and also plays an important role in fisheries management (Box 4.3).

In *Canada*, fish stock rebuilding may take place under its Fisheries Act which provides the legislative authority for the conservation of fish stocks and the management of fisheries, or the Species at Risk Act (SARA) which sets out specific requirements to protect and implement recovery strategies for all listed endangered or threatened species (terrestrial and aquatic).

Under *Japan's* Basic Law on the Fisheries Policy of 2001, the Government has been developing resource restoration plans and introducing a total allowable effort system for species that require urgent resource restoration (Fisheries Agency, 2007). A related framework for resource recovery plans to enable the implementation of the necessary measures in a comprehensive was established. Under this framework, national or regional levels of governments develop the resource recovery plans in co-operation with stakeholders.

**Box 4.3. An example of a coherent approach:
Rebuilding through endangered species legislation in Australia**

In some OECD countries, rebuilding fisheries may be achieved through or supported by national legislation regarding endangered species. These types of legislation, while geared more towards the protection of critically endangered species (whether they are terrestrial or aquatic) may also be triggered in some cases to allow for greater protection of depleted fisheries.

The Australian Fishery Management Authority (AFMA) manages fisheries under Commonwealth jurisdiction in accordance with the provisions of the Fisheries Management Act 1991. Australia also has an Environment Protection and Biodiversity Conservation Act 1999 (EPBC). There is clear guidance as to how the EPBC Act and the Fisheries Management Act interact with respect to the management of fisheries in terms of rebuilding, as follows.

- In situations where a stock biomass is determined to be above BLIM, it is not expected that the species would be added to the list of threatened species under the EPBC Act.

- In situations where the stock is at or below BLIM the risk to the species may be considered as unacceptably high; in these cases, stocks may be the subject to both the AFMA and the EPBC legislation.

- In cases where stock rebuilding strategy has been developed under the authority of AFMA and is in force, and if the termination of the strategy would negatively affect the conservation of the species, consideration would be given to listing the species in the conservation dependent category of the EPBC.

- If a particular stock falls significantly below BLIM, the guidelines note that there is an increased risk of irreversible impacts on the species. In these cases, the species will likely be considered for listing in a higher threat category which may trigger a requirement for the development of a formal recovery plan under the EPBC.

- For EPBC listed stocks where the biomass is above BLIM and is rebuilding towards BTARG, consideration may be given to removing the species from the EPBC Act list of threatened species, or amending the category it is placed in.

Source: DAFF (2007), Commonwealth Fisheries Harvest Strategy: Policies and Guidelines.

Related policy and guidance framework

Coherence across legislative and policy tools available for rebuilding is essential.

The level and extent of supporting policies and guidelines to support rebuilding legislation and plan design and implementation vary across respondents. New Zealand and Australia clearly articulate control rules, trigger points etc, through their respective Harvest Strategy Policies, as well as Canada through its Fisheries Act and SARA. The United States has developed detailed interpretation of their legislation. These supporting mechanism provide a degree of coherency across plans, and enhance transparency in their development. These are briefly described below, with additional details in the country chapters.

The Harvest Strategy Standard is a policy statement of best practice in relation to the setting of fishery and stock targets and limits for fish stocks in New Zealand's Quota Management System (QMS). It is intended to provide guidance as to how fisheries law will be applied in practice, by establishing a consistent and transparent framework for decision-making to achieve the objective of providing for utilisation of New Zealand's QMS species while ensuring sustainability. The stated objective is "to provide a consistent and transparent framework for setting fishery and stock targets and limits and associated fisheries management measures, so that there is a high probability of achieving targets, a very low probability of breaching limits, and acceptable probabilities of rebuilding stocks that nevertheless become depleted, in a timely manner. The Harvest Strategy Standard specifies appropriate probabilities that will achieve each of these outcomes." The Harvest Strategy Standard consists of three core components.

- A specified target about which a fishery or stock should fluctuate.

- A soft limit that triggers a requirement for a formal, time-constrained rebuilding plan.

- A hard limit below which fisheries should be considered for closure.

The Harvest Strategy further states that "use of a "soft" limit as a biological reference point that triggers a requirement for a formal, time-constrained rebuilding plan does not imply that no action needs to be taken to rebuild stocks that have fallen below targets but have not yet declined to the level of the soft limit. Management action needs to be continually applied to ensure that fisheries and stocks fluctuate around target levels, particularly when they start to fall below those targets. Such management action is likely to involve reductions in fishing mortality rates and TACs, and/or implementation or modification of input controls such as gear restrictions and seasonal or area closures. The role of the soft limit is to ensure that subsequent management action is sufficiently strengthened if previous action has not been adequate to prevent the stock declining to or below the soft limit".

The Harvest Strategy Standard is supported by Operational Guidelines, which consist of two key parts: (i) technical guidelines, which contain guidance on calculations of biological reference points to be used as inputs to setting fishing targets, and the basis for the default limits specified in the Harvest Strategy Standard; and (ii) implementation guidelines, which include sections on the transition period for implementing the Harvest Strategy Standard, the roles and responsibilities of science working groups and management working groups in estimating biological reference points and setting management targets, and the implications of implementing the Harvest Strategy Standard.

In Turkey, in addition to studies to align Turkish fisheries management with that of the European Union, preliminary fisheries plans have been prepared. The objectives include the rebuilding of depleted stocks, efficient resource management, introduction of fishing rights and sustainability of fishing opportunities for fishermen. Five preliminary fisheries management plans have been prepared, four of which are regionally-based (Black Sea, Marmara Sea, Aegean Sea and Mediterranean) while the fifth is based on fishing type (Inland Fisheries). Despite the principles and priorities set within certain national strategic documents, there is a need to develop a coherent fishery policy reflecting both the ecosystem approach to fisheries and the precautionary approach. These approaches should be better incorporated into national fisheries management in terms of conservation and sustainable management of fish stocks.

In the United States, NOAA Fisheries has revised the guidelines for National Standard 1 (NS1) of the Magnuson-Stevens Act, to comply with new annual catch limit and accountability measure requirements for ending overfishing in Federal fisheries as described in the Magnuson-Stevens Fishery Conservation and Management Reauthorisation Act of 2006. Specifically, the NS1 guidelines provide guidance on the timeline to prepare new rebuilding plans, Guidance on how to establish rebuilding time targets. The guidelines also provide advice on action to take at the end of a rebuilding period if a stock is not yet rebuilt and identifies two approaches for making overfishing status determinations. The NS 1 also sets out stock complexes may be formed for management purposes; this may be undertaken in cases where: stocks in a multispecies fishery cannot be targeted independent of one another and MSY cannot be defined on a stock-by-stock basis; there is insufficient data; or it is not feasible for fishermen to distinguish individual stocks among their catch.

The NS 1 also details how exceptions to the requirement to prevent overfishing may be dealt with, specifically the "mixed stock exception", and defines the circumstances under which overfishing of a stock in a mixed stock could occur. The guidelines state that the exception cannot be applied if a fishery is overfished. Before a council may recommend use of the exception to prevent overfishing, an analysis must be performed, and that analysis must contain a justification in terms of overall benefits, including a comparison of benefits under alternative management measures and an analysis of the risk of any stock or stock complex falling below its minimum stock size threshold.

In Canada, the Fisheries Act does not contain specifics concerning rebuilding. Rebuilding is done through Integrated Fisheries Management Plans which are under the legal authority of the Fisheries Act. If a stock begins to shows signs of decline then fishery management measures are to be introduced to reduce fishing mortality. For species listed as endangered or threatened a recovery must be developed within one or two years, respectively, of listing. To date no commercially harvested stocks have been listed under the SARA. In accordance with the precautionary approach, *A fishery decision-making framework incorporating the Precautionary Approach* (2009) states that healthy, cautious and critical stocks status zones should be based on defined upper stock and limit reference points. The policy requires that a rebuilding plan should be set up when stock abundance falls below its limit reference point. The rebuilding plan must include measures to limit fishing mortality and rebuild the stock to a level above the limit reference point in a timely fashion.

Structure and content of rebuilding plans

Planning the post-rebuilding management strategy early on provides certainty to stakeholders

This section provides an overview of rebuilding plans in various countries. Although such plans vary across countries they often include reference points, control rules and specific management measures. Some plans are also subject to regular reviews and evaluation in order to make course corrections as necessary. Post rebuilding management is often overlooked, but is nonetheless important so as to maintain a sustainable fishery that does not backslide or to avoid "boom and bust" cycles.

Multi-annual management plans are today the main tool used to rebuild EU fisheries. In particular, they are used to ensure that stocks are exploited within safe biological limits and that progressively the production of the stocks is maximised towards MSY. In some cases, where stocks are exploited outside safe biological limits, a recovery phase is initiated to bring them within safe biological limits. The main basis for this recovery phase is the precautionary approach and takes account of limit reference points recommended by relevant scientific bodies. Following a recovery, the plan enters the management phase that specifies the rules for inducing a progressive reduction in fishing mortality towards F_{msy}. Rebuilding plans in general include biological reference points, to identify the markers of "success" or "failure" as targets and warning points; rules for setting TACs as a function of current stock size estimates and fishing mortality rates; limits on TAC changes between years, applicable in some circumstances; and, effort management systems

The first plan which the European Union introduced for its own waters was the recovery of North Sea cod in 2004. Since then, the formula has been applied to a range of stocks in EU waters, and the Commission intends to progressively implement similar plans for all major commercial fish stocks over the coming years. Today, ten plans are implemented (North sea cod, North sea sole and plaice, Northern hake, Southern hake, Norway lobster, Bay of Biscay sole, Western channel sole, European eel, Baltic cod, West of Scotland herring) and another six are in the pipeline for 2009/10. A number of important international plans are also agreed with non-EU countries for stocks under joint management. Impact assessment is carried out during the design stage of multi-annual plans, which is a legal obligation before any policy proposal can be tabled. This includes extensive socio-economic analysis and stakeholder consultation, especially to examine the environmental, economic and social trade-offs between possible harvest control rules, and to establish the appropriate speed with which management measures towards F_{msy} should be taken. Socio-economic analysis relies on the use of bio-economic modelling and socio-economic data for EU fleets collected under the European Union's Data Collection Framework.

In the years immediately following the implementation of the moratoria in the early 1990s, Canada developed stock recovery strategies for depleted cod stocks in Atlantic Canada and Quebec. These recovery strategies were reviewed and adapted as needed following consideration of cod for possible listing under SARA, and three federal-provincial Action Teams were established in 2003: Canada-Newfoundland and Labrador; Canada-Quebec; and Canada-Maritimes. The Action Teams are mandated to develop integrated stock recovery and long-term management strategies for the cod stocks (see, for example, DFO 2005). The terms of reference for the Action Teams are focused on building an understanding of the current status of the cod stocks, increasing cooperation between stakeholders, and identifying and evaluating current science priorities and information with respect to management of the stocks. However, the Action Teams are not mandated to provide recommendations to the Minister of Fisheries and Oceans on cod stock management or annual TACs. Nor are they mandated to address access or historical share issues within the fisheries or to undertake new scientific research.

Specific to the precautionary approach and stock rebuilding, *A fishery decision-making framework incorporating the Precautionary Approach* (2009) for key commercial target stocks in Canada indicates that healthy, cautious and critical stock status zones are to be based on defined upper stock and limit reference point. This policy requires the implementation of a stock rebuilding plan when the abundance of a key stock falls below its limit reference point. The rebuilding plan must include measures to limit fishing mortality with the aim of rebuilding the stock above its limit reference point in a timely fashion. Canada is currently piloting a draft framework for incorporating socio-economic analysis into fisheries management plans with a view to explicitly dealing with the economic context around fisheries management decisions.

In January 2008, Australia implemented fisheries management changes in rebuilding plans for a number of overfished stocks. The Harvesting Strategy Policy (HSP) requires formal rebuilding strategies for all species that are below their biomass limit reference point. In 2008, formal rebuilding strategies developed for Eastern Gemfish and School Shark, both recently listed as Conservation Dependent species under the Environment Protection and Conservation Biodiversity Act 1999. A rebuilding strategy has been in place for Orange Roughy since 2006 when it was listed as Conservation Dependent.

In 2001, stock assessments suggested that the brown and grooved tiger prawns in Australia's Northern Prawn Fishery were depleted and fishing effort too high to promote recover to the fishery's target of MSY (the stock size required to achieve the maximum sustainable yield). The level of depletion of the prawn stock was not such that it would be regarded as "overfished" as the biomass had not gone below the HSP limit reference point. However, as it was below the fishery's target, a rebuilding programme was implemented and the most recent assessment suggests that the biomass of both species is around or above this target. In 2004, the fishery adopted Maximum Economic Yield (MEY) as the target. The primary fisheries management instruments are input controls such as limited entry, seasonal and area closures, number of fishing vessels, and gear and mesh size restrictions. Output controls such as individual transferable quotas as part of a total allowable catch are also used in many of the fisheries and many fisheries are moving to this form of management.

Under the Harvest Strategy Standard in New Zealand, there is a requirement for the development of formal stock rebuilding plans for stocks that have breached the soft limit. The default soft limit is ½ B_{MSY} or 20% B_0, whichever is higher. Stocks that have fallen below the soft limit should be rebuilt back to at least the target level in a time frame between T_{min} and $2 * T_{min}$ with an acceptable probability. Stocks will be considered to have been fully rebuilt when it can be demonstrated that there is at least a 70% probability that the target has been achieved[1] and there is at least a 50% probability that the stock is above the soft limit. The hard limit is the biological reference point at which closure should be considered for target fisheries; it may be also be appropriate to consider curtailment or closure of fisheries that incidentally catch the species concerned. The default hard limit is ¼ B_{MSY} or 10% B_0, whichever is higher. At this point in time, there is no required structure for the formulation or documentation of rebuilding plans beyond these guidelines, which are also elaborated to a certain degree in the current version of the Operational Guidelines.

In Norway, the composition of a rebuilding plan depends on the objectives put forward as regards time span of recovery, fishing activity during the rebuilding phase, levels of risks, level of assessment during rebuilding, etc. In most circumstances, a total ban on a fishery to let a fish stock recover is not introduced. A step-by-step reduction in fishing mortality is the most frequently used tool to rebuild. The initial challenge is to reduce harvest in the first stages of the rebuilding plan because this necessarily entails a significant reduction in income for stake holders in the short term. However, some species have recovered rapidly when fishing pressure has been reduced, and Norwegian fishers have experienced the advantages from investment in future yields, also at an individual level.

In the United States, rebuilding strategies generally are incorporated into existing fisheries management plans (FMPs) as an amendment. If an FMP does not exist, the legislation requires that one be developed within one year. Any stock that has previously been listed, or is currently listed, as overfished is required to have a rebuilding programme until the stock has been rebuilt to levels consistent with supporting MSY on a sustainable basis. Many of the stocks listed as overfished have experienced excessive levels of fishing effort in recent years, while other stocks may be listed as overfished because of prevailing environmental conditions, habitat degradation, or natural fluctuations in the stocks. These factors may have reduced the stock biomass to levels below that necessary to produce MSY on a continuing basis. Sometimes, management measures have little impact on the status of the stocks. For example, many of the Pacific salmon stocks under the Pacific Fisheries Management Council jurisdiction are not significantly impacted in fisheries within the Council's jurisdiction. Other stocks are listed as threatened or endangered under the Endangered Species Act, and management for these stocks is conducted under the Act.

Review and evaluation processes

As expressed in Council Regulation 2371/2002 (Articles 5 and 6), "the Commission shall report on the effectiveness of the management plans in achieving the targets." The 2008 "Impact Assessment Regarding the Commission's proposal establishing revised measures for the recovery of cod stocks" recognised that the initial strategy to rebuild cod stocks was insufficient and did not meet its intended goals. In the development of an updated rebuilding plan, the European Union revised long-term objectives *from* targeting specific biomass measures *towards* an approach of striving to achieve an optimum exploitation rate to lead to the highest sustainable yield. In response to comments from member states, the complexity of the management system was reduced while at the same time bestowing flexibility for countries in terms of implementation. Finally, the updated plan should also deal more directly with the issue of discards and benefit from the development of clearer harvest rules so as to lessen the amount of impromptu decisions.

Following the request of the Commission, the Scientific, Technical and Economic Committee for Fisheries (STECF) has outlined a proposal for a framework to undertake future evaluations of existing plans, in particular for North Sea sole and plaice, Baltic cod, hake, nephrops, Bay of Biscay sole and Western channel sole in 2009-10. The format of this evaluation is as follows.

- Background Information and data – historical background, objectives and reference points, provide up to date fishery/fleet data

- Elements to be reviewed

    o Implementation: design issues, enforcement and compliance

    o Environmental effects: fishery and stock response and impact of management measures, consistency of targets and reference points

    o Ecosystem effects: discarding practices, by-catch rates, habitat degradation

- Social and economic effects - data and calculation of indicators for fleets concerned (or general socio-economic CFP objectives if no specific objectives are defined), cost effectiveness (e.g. do the benefits outweigh the cost of implementation and enforcement).

- Added value of the plan - what is likely to have happened if the management plan had not been put in place, costs/benefits of plan in environmental and socio-economic terms.

- Performance evaluation of the plan (based on the above) – effectiveness, utility, efficiency (cost-effectiveness), appropriateness of indicators, sustainability of plan

- Conclusions – global judgment of plan and recommendations for future revisions and evaluations (data, indicators, objectives)

In New Zealand, once a rebuilding plan has been implemented, Science Working Groups will regularly evaluate and report on the performance of the rebuilding plans. To the extent practical, this will be based on regular monitoring programmes or stock assessments.

In Norway, the assessment of the stocks which are subject to a rebuilding programme is of essential importance. How a fish stock responds to catch restrictions may only be estimated roughly, but an increased number of previous programmes and the data derived from them may contribute to narrow the predicted intervals of estimated effects from future programmes. The assessment implies steady financial contributions to research groups which monitor the development and collect as much data as possible, but also careful monitoring of the activity of the fishing fleet and collection of comprehensive catch statistics. If the estimated effects fail to appear, one may need to reconsider the approach by applying more restrict harvest control rules. This can be carried out, by example, taking larger annual steps in the progressive reduction of fishing mortality. The balance between short-term burden for stakeholders and the importance of rapid stock improvement is an important issue for consideration.

Post-rebuilding management

A key, but often neglected, issue in the design and implementation of fisheries rebuilding plans is the management of the fishery once the stock is rebuilt to the target level. This is important for two reasons. First, failure to resolve any underlying management issues that contributed to the need to rebuild the fishery will lead to these resurfacing again once any rebuilding measures (such as catch, effort or gear restrictions) are removed. The period of rebuilding may be regarded as requiring special measures that may be seen as dramatic and time bound by fishers (and managers). There is a need to ensure that the post-rebuilding management measures alter the fundamental incentives facing fishers are appropriate and do not encourage the excess capacity or effort in the future.

Second, the discussion and agreement on post-rebuilding management amongst stakeholders at the design stage will provide a greater degree of certainty for stakeholders. This, in turn, will likely ensure a higher level of support for the entire rebuilding package, reducing the risk of default or backsliding by participants, and potentially reducing the costs of enforcement. This may also require a significant up-front investment in negotiating and planning the parameters of the rebuilding process.

There are currently no specific policies or guidelines associated with post-rebuilding and maintenance approaches at EC level. In New Zealand, by definition, well-managed fisheries are those that fluctuate around appropriate targets and remain well above limits. Management action should ensure that this situation continues. The Operational Guidelines specify the types of management actions that should be used to ensure that fisheries fluctuate around appropriate targets, well above limits. In Norway, when stocks that were once overfished have recovered, and the rebuilding phase is accomplished, it can be questioned how to manage the stock in the continuation. The most important objective will be to maintain the spawning stock biomass at a safe and reproductive level. However, there are optional strategies which may all comply with this requirement. At this stage, it may be of interest to include new objectives which take into consideration social and socioeconomic aspects.

In Australia, the issue of when to allow targeted fishing after a stock recovers to above B_{LIM} is also considered in the HSP. For stocks that have recovered from below B_{LIM}, and have not been listed in vulnerable or a higher threat category, targeted fishing will be allowed as long as fishing does not interfere with the agreed stock rebuilding strategy, as agreed to by the AFMA and the environment minister. For stocks that have been listed as threatened under the EPBC Act, where the biomass of a listed stock is above B_{LIM} and rebuilding towards B_{TARG}, the HSP states that consideration could be given to removing the species from the threatened species, or amending the category it is in.

Stakeholder consultation and engagement

Stakeholder involvement is key for successful rebuilding

Most government policies and programmes now explicitly require that stakeholders be engaged in the development of rebuilding plans from the initial stages to allow for the broadest level of support. This may extend to stakeholder involvement in the provision of biological and economic data, participation in the development of rebuilding targets and paths, as well as the analysis of social and economic impacts, and the subsequent implementation of the plan.

Stakeholder involvement in the design of rebuilding plans is particularly important to ensure that all the best available information is used in the decision making process. In addition, the degree of uncertainty in scientific assessments should be clearly articulated to stakeholders, given that many decisions will be made with less than perfect information. Obtaining the agreement of fishers on a set of pre-established harvest control rules as the status of the stock changes in response to rebuilding measures is of the utmost importance. This will serve to limit the calls for increasing quotas or removing restrictions when there are signs of improvement instances as well as demonstrate that adaptive management is the approach that is being pursued in the face of varying levels of uncertainty about the stocks response to rebuilding measures, and external factors such as climate change (Caddy and Agnew, 2004).

One characteristic of the Korean ecosystem based Fish Stock Rebuilding Plan (FSRP) is the active involvement of various stakeholders which is encouraged in the whole process of development, implementation and assessment of the FSRP. Stakeholders include, for example, fishermen, people from academia, government officials and researchers. An important aspect is the premise of voluntary participation of fishermen and other stakeholders connecting with community-based management fisheries. The participation of fishermen is important for the rebuilding plans to be efficiently carried out. To take an example, the voluntary agreements by fishermen concern decisions concerning gear, number of fishing days and choice of protected areas.

In the European Union, the EC considers that real dialogue is a prerequisite for successful policies as it generates an exchange of views with fishermen and other stakeholders and provides the Commission with better knowledge about their problems and expectations which in turn can be taken into consideration when proposals for fisheries rules are drafted by the Commission. Stakeholders in the fishing industry are also more likely to accept and implement CFP rules if they have been involved in the formulation of these rules.

The EC has taken a series of measures to strengthen the dialogue with the fisheries sector and other interested parties. One of the first measures was to set up the Advisory Committee on Fisheries at the beginning of the 1970s. The Committee was reformed in 2000 to make it more efficient and to broaden the dialogue with the industry and other stakeholders. New interest groups (aquaculture, NGOs and scientists) became involved in the committee which became known as the Advisory Committee on Fisheries and Aquaculture (ACFA) which is consulted by the Commission on measures related to the CFP and can issue opinions on its own initiative.

Despite the progress achieved in terms of strengthening the dialogue with stakeholders, the consultation of the fishing industry in the framework of the 2002 reform of the CFP clearly showed there was a need to do more. Stakeholders did not feel sufficiently involved in several important aspects of the CFP, such as, for example, the provision of scientific advice and the adoption of technical measures. Many fishers, in particular, believed that their views and knowledge were not sufficiently taken into account by managers and scientists. To address this shortcoming, the Commission proposed a network of Regional Advisory Councils (RACs) involving fishers, scientists and other stakeholders on a regional level. On the basis of the Commission proposal, the Council adopted in July 2004 a common framework for RACs which foresaw the establishment of seven RACs covering 5 geographical areas as well as pelagic stocks and the highs seas fleet. They now enable the fishing sector to work more closely with scientists in collating reliable data and discussing ways of improving scientific advice. RACs submit recommendations and suggestions on any aspects of the fisheries they cover to the Commission and the member states concerned. Each multi-annual plan is drawn up in close collaboration with the RAC concerned, both for the technical content and the evaluation of the socio-economic impact.

In New Zealand, the need to implement a rebuilding plan is often identified by a Science Working Group, usually a Fisheries Assessment Working Groups (FAWG). Fisheries managers are then responsible for developing rebuilding plans that fit the parameters of the Harvest Strategy Standard, in consultation with Maori, stakeholders and scientists. Since the Harvest Strategy Standard allows a time frame for rebuilding between T_{min} and twice T_{min}, it is essential to engage with stakeholders and Maori[2] and stakeholders to determine the social and economic constraints under which they operate. Any non-emergency change to a TAC or TACC requires the development of, and consultation on, an initial position paper under Section 12 of the Fisheries Act. A final advice paper is prepared on the basis of feedback from Maori, environmental, recreational and commercial interests.

Under the Magnuson-Stevens Act and other statutes in the United States, there is a comprehensive process for ensuring participation by stakeholders, including commercial and recreational users, and other interested constituencies. Consultation and engagement with all these stakeholders is provided for in detail in the Magnuson-Stevens Act, especially in Section 302, which addresses the operations of the regional fishery management councils. Essentially, commercial and recreational users and other constituent groups may participate in the management process through like-minded Council members, by attending public meetings, and providing testimony and comments on proposed actions. In addition, all these constituent groups may participate through the political process by contacting their elected Congressional representatives.

In Canada, in cases where it is believed that impacts from listing fish stocks at-risk are high, the Canadian *Cabinet Directive on Streamlining Regulation* requires the Minister to conduct public consultations and socio-economic analysis to inform the decision. Additional consultations are needed where Aboriginal or First Nation land claims and treaty rights are concerned. If species are listed on SARA the considerable discretion in management fisheries under the Fisheries Act is reduced. Harvest decisions are likely to be guided by automatic prohibitions of the Act unless scientific assessments determine that some level of harvest will not jeopardise the recovery of the species. However, when managing commercially harvested stocks, economic considerations prevail primarily in situations where the specific stock is considered to be in the healthy zone, but neither in the critical nor cautious zones.

In Norway, the involvement of stakeholders in management decisions in the annual regulations is achieved through the Advisory Meeting for Fisheries Regulations, which is a public and open meeting. In the meeting fishers associations, the fishing industries, trade unions, the Sami Parliament, local authorities, environmental organizations and other stakeholders are represented and can express their opinions. The Regulatory meeting is the main tool to secure involvement and participation of stakeholders. During the year stakeholders are involved in other management decisions though public hearings and consultations. Parts of the fisheries are subject to regulations in some form or other, and the regulatory arrangements deal with conservation as well as allocation issues. The domestic fisheries management regime employs three sets of regulatory instruments, as well as a comprehensive enforcement scheme. As already explained in the section dealing with rebuilding plans, the stakeholders are involved through a consultation process when new rebuilding plans are to be introduced. Then the balancing of short term burden to the stakeholders, and the importance of rapid stock improvement, must be considered.

Australia (AFMA) places emphasis on a partnership approach between fisheries managers, industry, scientists, fishing operators, environmentalists/conservationists, recreational interests and the general public. Implementation of the partnership approach is facilitated by management advisory committees (MACs) or consultative committees, which have been established for all major Commonwealth managed fisheries[3]. The MACs typically consist of the AFMA manager for the fishery, industry representatives, a research scientist, a conservation member and, where relevant, a member representing state or territory governments and a recreational fishery or charter boat fishery representative. Consultative committees have a similar structure but apply to smaller or developing fisheries. Resource assessment groups provide assessments of the status of target, by-product and by-catch species, and assessment of the broader marine ecosystem to both types of committee.

Financial or other instruments used to support rebuilding

Transition mechanisms are important to obtain and maintain support for reforms implemented as part of rebuilding

The European Commission, Norway and Australia recently dedicated funds or developed broad mechanisms, such as buyback programmes, to accompany rebuilding programmes. Other countries often have the necessary legislative and regulatory framework to set-up such programmes should it be considered necessary in a particular case. The United States, for example, can create buyback programmes through the Magnusson-Stevens Act, and have setup such a programme since 2007. In New Zealand, compensation programmes have not been used since 1986.

The European Union has developed a financial instrument for the transition towards more sustainable fishing. The European Fisheries Fund allows member states to finance the gradual restructuring of the sector which currently suffers from overcapacity and includes *inter alia* vessel decommissioning and development of alternative economic activities outside the fisheries. In the context of fisheries rebuilding, there are also financial incentives for vessels to improve gear selectivity. The European Fisheries Fund (EFF) is planned to run for seven years (2007-2013) with a total budget of around EUR 3.8 billion. This is seen as the main alternative approach to the more extensive use of rights-based management (RBM). The member states choose the projects that are granted EFF co-funding.

In New Zealand, no compensation programmes have been applied since the inception of the quota management system in 1986. There is broad agreement in New Zealand that the commercial fishing industry will not be subsidised.

In Norway, a fund to decommission fishing vessels up to 15 metres holding annual permit(s) was established on 1 July 2003. The scheme was partly funded through a fee on the landed value of every Norwegian fishing vessel. The Government has so far transferred NOK 108,25 million to the fund, estimated to be about 50% of the industry's contribution. The programme was initially terminated 1 July 2008, however, due to remaining funds, the programme was extended to 2009. As licenses of the scrapped vessels are withdrawn and redistributed to the remaining vessels, the aim of the fund is to improve the profitability of the remaining vessels. At present, there are no economic instruments used directly to support fisheries rebuilding because of the continuous increasing productivity of the industry, Norwegian fisheries authorities must at all times consider market-based measures to ensure profitability and a sustainable utilisation of the resources. Fewer vessels and fishermen are inevitable in the future, and subsidies will only delay the transition.

Canada has used a number of retraining programmes to ease the transition of individuals out of the fishery and thereby decrease pressure on stocks. Temporary income replacement programmes have also been used to support people and communities during downturns in the fishery sector.

Although there is no formally designated programme to provide economic assistance to facilitate the adoption and administration of rebuilding plans in the United States, limited assistance may be provided in specific circumstances. One example is a fishing capacity reduction programme, as provided for in Section 312 (b-e) of the Magnuson-Stevens Act. With this approach, a buyback programme reduces harvest capacity and may facilitate the industry's support for the restrictions associated with a rebuilding programme. NOAA Fisheries financial services experts work with fishing vessel owners to develop and implement these programmes. In addition, some limited assistance may under some circumstances be provided to fishermen who operate in rebuilding fisheries. For example, the Magnuson-Stevens Act 2007 includes a newly authorised programme, the Fisheries Conservation and Management Fund, which may be used to, inter alia, for "providing financial assistance to fishermen to offset the costs of modifying fishing practices and gear to meet the requirements" of this law.

Australian government fisheries are managed under systems designed to allow adjustment to occur autonomously through market processes; this includes adjustment in response to commercial and biological conditions. The government does not provide financial support for fisheries rebuilding, including with respect to waiving of management fees, which are cost-recovered. However, in 2006-2007 the Australian Government implemented the Securing our Fishing Future structural adjustment package to address environmental (state of Australian fish stocks) and economic concerns (sustainable and profitable industry). The package included AUD 149 million for a one-time capped fishing decommissioning scheme that focused on reducing the high level of fishing capacity in fisheries subject to over-fishing, or at significant risk of over-fishing in the future. This component also addressed the displaced fishing effort arising from the creation of Marine Protected Areas in the South east Marine Region.

In Korea, measures to support fishers during rebuilding periods are being considered in order to encourage voluntary participation in the plans (for example, supports for reduction in fishing effort such as limitation on the number of fishing days and suspension system, improvement of fishing grounds for selective fishing of small sized fishes and avoidance of mixed fishing, aids for expenses on disposition of fishing gears, and support system on training of fishers).

Addressing economic and social aspects

There is increasing recognition that considering economic factors early in the planning process can increase the likelihood that rebuilding will be successful

Countries such as the United States and New Zealand have legislative requirements to consider socio-economic issues when rebuilding or managing stocks. The action of rebuilding may also involve introducing new regulations which can also trigger various economic assessments. Challenges related to the integration of socio-economic considerations include the lack of data and/or data compatible with biological data. Socio-economic data is becoming more reliable however and countries are increasingly incorporating socio-economic impact analysis (e.g. revenue, costs, gross value added, profits, and employment) at the early stage of designing rebuilding plans. Market based measures are also being explored in several countries as part of rebuilding

In the European Union, the primary way by which economic analysis is built into rebuilding plans is through STECF evaluations and Impact Assessments. As economic and social data become more reliable and compatible with biological data, socio-economic analysis is beginning to play a bigger role in the design of plans. A recent study examines all the available bio-economic models that can be used for a range of fisheries management purposes[4]. Although the progress in this area is steady, the incorporation of socio-economic data (e.g. revenue, costs, gross value added, profits, and employment) and analysis at the early design stage of plans is limited.

In New Zealand, the 1996 Fisheries Act requires that relevant economic, social and cultural factors be taken into account in deciding upon the way and rate at which a stock is rebuilt to the target level, towards or above MSY.

- In the case of stocks with significant allocations to more than one sector (greater than about 20% of the TAC), there may be considerable disagreement about timeframes for rebuilding. Where a stock is virtually exclusively allocated to one sector, the timeframe selected may be more reflective of the interests of that particular sector. Section 13(3) of the Fisheries Act (1996) states that "in considering the way and rate at which a stock is moved towards to or above a level that can produce the maximum sustainable yield … the Minister shall have regard to such social, cultural, and economic factors as he or she considers relevant". The Fisheries Act requires that relevant economic, social and cultural factors be taken into account in deciding upon the way and rate at which a stock is rebuilt to the target level.

- Economic inputs can be used to set target reference points and recovery timeframes under the Harvest Strategy Standard, provided these meet or exceed relevant elements of the Standard. Economic impacts of closures or other rebuilding measures in the case of mixed fisheries are also acknowledged: in this case, action should be taken early on with fishers to create proper incentives.

In the United States, all significant fisheries conservation and management measures, including rebuilding plans, must include certain economic assessments. These regulatory assessments are required by the Magnuson-Stevens Act, other laws, and a few Executive Orders. The major economic impacts and issues that must be considered are as follows:

- Magnuson-Stevens Act: economic efficiency, no excessive shares, cost minimisation, cumulative economic impacts, evaluation of economic impacts and recovery benefits on commercial, recreational, and charter sectors.

- National Environmental Policy Act: impacts on human environment.

- Regulatory Flexibility Act: impacts on small entities.

- Executive Order 12866: assessment of net benefits.

In addition, the Magnuson-Stevens Act includes discretionary provisions that may apply to the assessment of economic impacts of rebuilding plans. For example, if a rebuilding plan includes a limited access system, section 303(b)(6) on limited entry requires examination of "(A) present participation in the fishery, (B) historical fishing practices in, and dependence on, the fishery, (C) the economics of the fishery, (D) the capability of fishing vessels used in the fishery to engage in other fisheries, (E) the cultural and social framework relevant to the fishery and any affected fishing communities, and (F) any other relevant considerations."

In Korea, bio-economic modelling is used for analyzing economic impact changes and achieving the target stock based on stock assessment for each species. In particular, biological and economic uncertainties are fully considered as part of the bio-economic modelling. It is used for selecting effective stock rebuilding measures on the basis of impact assessment for various fishery management measures.

In Canada, the economic analysis required by SARA comes into play at the recovery action planning stage for endangered or threatened species. Those action plans must contain "an evaluation of the socio-economic costs of the action plan and the benefits to be derived from its implementation". Under the Fisheries Act there are no formal requirements regarding economic objectives for fisheries management. Economic analysis is therefore done on an ad hoc basis. However, Canada has recently begun to include economic information within IFMPs, which will enhance capacity for the use of economic information and incentives for rebuilding.

Application of market based measures or incentives

In the European Union, the area of market based measures or incentives remain the competence of the member states. This is particularly true for the possible application of rights-based management systems (RBM). A recent study in 2009 gives an overview of RBM application in EU member states (*ec.europa.eu/fisheries/publications/studies/rbm_2009_part1.pdf*). The study has identified 63 different RBM systems in marine fisheries. Of these, 47 (75%) are classified as having weak property rights, three fisheries (5%) have strong non-transferable property rights, and 13 fisheries (20%) have RBM systems with strong tradable rights, the latter found in Spain, Portugal, The Netherlands, Denmark, Estonia and the United Kingdom. However, the development in this area is still rather modest and *ad hoc*. The potential role of RBM in EU fisheries management is being discussed in the current CFP reform consultation, especially in relation to how RBM can be used to help restructure the fishing fleet and reduce overcapacity, and give incentives for greater stewardship. This is deemed especially important in the context of fisheries rebuilding.

In the United States, market based programmes are authorised by Section 303A of the Magnuson-Stevens Act, which addresses limited access privileges programmes

(LAPPs). A LAPP programme is one means of reducing fishing mortality and overcapacity, and therefore may be a part of a rebuilding programme.

The Magnuson-Stevens Act 2007 authorises that a Council may submit, and the Secretary may approve, for a fishery that is managed under a limited access system, a limited access privilege programme to harvest fish if the programme meets the requirements of the Act. Specifically, the term "limited access privilege" programme (LAPP): (A) refers to a Federal permit, issued as part of a limited access system under Section 303A to harvest a quantity of fish expressed by a unit or units representing a portion of the total allowable catch of the fishery that may be received or held for exclusive use by a person; and (B) includes an individual fishing quota; but (C) does not include community development quotas as described in section 305(i).

To help the councils develop and implement these and similar programmes, NOAA Fisheries is presently working on regulatory guidelines on various LAPP issues and a broader policy on "catch share" programmes. Catch share programmes is a designation that includes LAPPs and possibly other management programmes that allocate harvest shares exclusively to designated recipients. With a policy and strategy on catch share programmes in place, the councils and the industry groups they represent will have a variety of exclusive allocation programmes to choose from, including individual fishing quotas, community quota programmes, fishing cooperatives, and other programmes such as sector-specific allocations.

Key insights

The country profiles developed for this report provide an overview of the fisheries rebuilding process in certain OECD countries and several RFMOs. Although the approach, policy framework and legislative basis for fisheries rebuilding varies across countries, there are common insights that can be drawn and generalised to the methods used in terms of developing rebuilding plans.

Charting progress of national rebuilding plans and stock status improves transparency and can help articulate the benefits of the plans. National assessments provide a transparent picture of how rebuilding plans are progressing and enable progress to be measured. There is also the fact that being transparent on the progress of rebuilding improves communication and increases credibility; it allows both stakeholders and the general public to get a clear picture of stock trends and could help increase support and acceptance of rebuilding measures. It is also important to publicise the status and progress of rebuilding plans in order to make information not only available but known to interested parties.

Legislative and regulatory support provides a strong foundation for rebuilding, while policies and guidelines provide more detailed direction. A clear legislative requirement to rebuild fisheries has been demonstrated to be a factor in successful rebuilding plans by communicating clearly the importance of sustainable and viable fisheries by making rebuilding a legal duty. Supporting policies and guidelines provide a flexible environment that articulates how the legal requirements will be adhered to and how rebuilding will be undertaken. In the countries surveyed, these policies and guidelines often specify harvest control rules, direction about setting targets and about how to deal with risk and uncertainty. Supporting policies enable the gathering and sharing of information to occur in an open and transparent manner, where the "rules of the game" are defined at the outset; this also promotes public trust.

Coherence across legislative and policy tools available for rebuilding is essential

In many countries, more than one legislative tool is available to support rebuilding. Well defined triggers that indicate which tool is most appropriate under which circumstance goes a long way in building trust with stakeholders. Incoherence can undermine policy objectives and be counterproductive; coherent policies are mutually reinforcing.

Planning the post-rebuilding management strategy early on provides certainty to stakeholders

Understanding how the fishery will be restructured after rebuilding enables the development of plans and supporting measures to reach their goals.

Stakeholder involvement is key for success

Countries surveyed have emphasised the importance of collaborative processes that engage stakeholders and partners from the design of rebuilding plans through to the implementation process. Stakeholder involvement is important because it increases the likelihood that measures will be accepted and supported by stakeholders and facilitates implementation, acceptance and uptake. Stakeholder involvement leads to the development of a joint plan and ultimately shared responsibility.

Transition mechanisms are important to obtain and maintain support for reforms made as part of rebuilding

A major challenge in rebuilding is to address the issue of the distribution of the impacts resulting from implementing rebuilding measures and maintaining management changes that have been implemented. In this regard, some countries have developed compensation programmes as part of the rebuilding plan to build and maintain support from affected stakeholders and groups who may be otherwise very vocal.

There is increasing recognition that considering economic factors early in the planning process can increase the likelihood that rebuilding will be successful. Robust economic analyses that illustrate the impact of various measures are becoming increasingly utilised by governments early in the process. In some countries, these economic analyses are mandated in various legislations.

Notes

1.   Use of a probability level greater than 50% ensures that rebuilding plans are not abandoned too soon. In addition, for a stock that has been depleted below the soft limit, there is a need to rebuild the age structure as well as the biomass, and this may not be achieved by using a probability as low as 50%.

2.   Maori are treated as partners, rather than stakeholders, and therefore have a different status. They are generally not referred to as stakeholders.

3.   The only exceptions are the Coral Sea and South Tasman Rise Fisheries.

4.   For further information, see *ec.europa.eu/fisheries/publications/studies/bio-economic_models_en.pdf.*

References

Caddy, J.F. and D.J. Agnew (2004), "An Overview of Recent Global Experience with Recovery Plans for Depleted Marine Resources and Suggested Guidelines for Recovery Planning", *Review of Fish and Fisheries*, Vol. 14, pp. 43-112.

Japan Fisheries Agency (2007), *Japan's Fisheries*, Tokyo, September, available at *www.jfa.maff.go.jp/*.

Larkin, S., Sylvia, G., Harte, M., Quigley, K. (2007). "Optimal Rebuilding of Fish Stocks in Different Nations: Bioeconomic Lessons for Regulators". *South Atlantic Fishery Management Council Marine Resource Economics*, Volume 21, pp. 395–413.

Wakeford, R.C., D.J. Agnew and C.C. Mees (2007), *Review of Institutional Arrangements and Evaluation of Factors Associated with Successful Stock Recovery Programs*, CEC 6th Framework Programme No. 022717 UNCOVER, MRAG Report, March.

Wiedenmann, J. and M. Mangel (2006), *A Review of Rebuilding Plans for Overfished Stocks in the United States: Identifying Situations of Special Concern*, Lenfest Ocean Program, Washington DC, June, available at *www.lenfestocean.org\publications*.

WWF (n.d.), *Essential Guide to Successful Recovery Plans for Europe's Fish Stocks*, WWF Europe, *www.panda.org/marine*.

Chapter 5.

Making reforms happen in fisheries

This study shows that rebuilding fisheries often calls for reforms in fisheries policies and sometimes a change in the fisheries management framework. Previous OECD work has provided avenues for successful reforms, including the necessity to obtain stakeholders agreement on the status of the fishery and the objectives of the rebuilding plan. Stakeholder involvement is crucial as they may provide important inputs into the reform process; for example, by providing information about risks and uncertainties. The objectives of the reforms must be realistic and attainable. If not, the reforms will lack credibility and result in a low chance of success. OECD work shows that rights-based fisheries management measures have often been successful in rebuilding fisheries. Such rights can create the incentives for stakeholders to have vested interests in rebuilding fisheries.

What is needed for successful reform? The fact that a fishery is in need of rebuilding indicates that the management regime in place has not been successful and that reforms are needed. Decisions regarding whether and how to rebuild a fishery are policy choices where various factors affect the decisions made. Such policy choices reflect social, economic, environmental and political realities.

Although considerable benefits can be made from rebuilding, and in many cases inaction is not an option, implementing the necessary forms can be a challenge. Considerable work has already been done at OECD on how to successfully bring about reforms. A recent publication gives an account of national experiences in fisheries policies reform (OECD, 2011) while discussions on the political economy issues of such reforms in fisheries is given by Sutinen (2008) (Box 5.1).

A rebuilding plan is not just a technical issue, but one that requires specific actions by all stakeholders involved. A plan, no matter how well designed in technical terms, will fail if stakeholders do not adhere to it and act accordingly. One way to gain stakeholder support, or at least reduce their opposition, is to involve them in the process of developing and implementing the plan especially insofar as their involvement can bring specific knowledge about the fishery.

Box 5.1. The political economy of reform

Despite 30 years of fishery management programmes, most coastal nations have not yet succeeded in effectively controlling activities in their waters, or maintaining healthy fish stocks. It has been estimated that in 2005, half of marine fish stocks were fully exploited and about one quarter of stocks was overexploited, depleted or recovering from depletion. Nonetheless, it is not uncommon that statistical data and scientific evidence are ignored in policy setting. For example, total allowable catch rates are frequently set above the rates recommended by fishery scientists as necessary for sustainability. Reasons for such "governance failures" include: i) special interest effects; ii) rational voter ignorance; iii) bundling of issues; iv) short-sightedness; v) decoupling of costs and benefits; and vi) bureaucratic inefficiencies.

Short-sightedness of the principal actors and decoupled benefits and costs of fishery products have a powerful influence on the choice of fishery management policies. Politicians often exhibit short-sightedness by enacting special legislation and appropriations for fisheries. Fishermen, in turn, tend to be short-sighted because they have no secure claim on future outcomes in their fishery, and because of the great uncertainty about future fishery policies, fish stocks and markets. Thus, effective conservation policies tend to be disfavoured because they concentrate short-term costs upon resource users in exchange for benefits in the future that would not necessarily accrue to the users who make the sacrifice.

Only when those who sacrifice in the present can expect to receive benefits in the future can the political marketplace of fishery be expected to produce effective conservation policies. To correct or minimise governance failures in fisheries, national administrations can introduce strong property rights (e.g. transferable individual licences and individual quotas), decentralise rights and responsibilities to individuals and user groups and implement cost recovery and various forms of sustainable financing mechanisms in order to change the incentive structure. Ultimately, the success of any of these measures depends on the interests of and support from private sector actors.

Source: Sutinen (2008).

Determining and agreeing on status and objectives

Although stakeholder involvement is situation-specific and may vary there are two issues that must be decided on before designing a specific plan of action to rebuild a fishery, i.e. what is the state of the fishery and what are the objectives of the plan.

An important step is to evaluate the state of the fishery with regards to biological, industrial and societal characteristics, and to determine why the fishery is facing challenges. Major problems emerge when stakeholders do not agree at least broadly on these issues. Whether the fishery requires rebuilding because of overfishing, ecological changes, and/or other factors may also be the subject of disagreement, and the design of the rebuilding plan will be heavily influenced by the evaluation of these causes and how they are addressed.

Uncertainties regarding biological and economic data may cast doubt on the status of stocks and fisheries which again can lead to a lack of agreement on the state of the fishery and whether and how it should be rebuilt. Under such circumstances, stakeholder involvement in data collection and their perception of risks and uncertainties might be helpful given their knowledge and experience. A consensus on status must, however, be consistent with scientific evidence, within reasonable bounds determined by scientific uncertainty and rigorous scientific practice.

Reaching an agreement on the state of the fishery is often problematic in shared fisheries which are outside of exclusive national management areas rendering effective rebuilding plans and enforcement challenging. This is reflected in many of the case studies on RFMOs, carried out for this project. The case of Greenland Halibut in the North Atlantic highlights some of the difficulties to manage shared stocks in international waters where the absence of an agreement between governments and stakeholders on the state of the fishery had a negative effect on rebuilding efforts. In 2003, NAFO member countries agreed on a 15-year rebuilding plan using TACs distributed among member states. Considerable resources were allocated to the management of this fishery with little success. Fishing mortality was higher than envisaged by the plan with catches consistently above the set TAC, and the management plan did not consider economic issues. There was widespread discord with regard to the scientific evidence and although various stakeholders had a voice in the process they had no formal decision power. The situation has changed since then, but this past history casts a light of some of the common challenges of managing international fisheries.

Another case is the Southern Bluefin Tuna (SBT) which is managed by the Commission for the Conservation of the Southern Bluefin Tuna (CCSBT). The CCSBT includes six members and three co-operating non-members (including the European Union). They have not developed a specific rebuilding plan, but their initial aim (late 1980s) was to rebuild by 2010 the parental biomass to 1980 levels. It became clear, however, during the "rebuilding phase" that this objective would not be met so the targets were changed accordingly. In (2009) an objective was set to reach 20% of the original spawning stock but with no target deadline. Indeed, it had been difficult to reach a consensus on the TAC from 1997 to 2003, and there was disaccord on the scientific evidence presented, recruitment was poor, and management arrangements did not include all the nations that harvested the stock. The focus had been to achieve biological targets while economic considerations were not high on the agenda, leading to a short-term strategy of achieving an (annual) increase in the parental biomass and reducing the risk of a recruitment decline. The main management measures used were a global TAC, national allocations, an approved vessel list, and a Trade Information Scheme (TIS). All this proved to be of little use, resulting in constant overfishing and declining biomass. Although a considerable amount of money was used in the scientific management of this stock, no specific economic analysis was undertaken by the CCSBT in the design of the rebuilding programme.

Agreeing on and setting objectives is an important component of rebuilding efforts. However, while an unsustainable fishery is usually socially suboptimal, different stakeholders and members of society are likely to differ with respect to the values that guide the rebuilding strategy. Different stakeholder groups are therefore likely to have conflicting views on suitable objectives as well as on the actions to be taken and by whom, with each group preferring objectives and measures that match their own preferences. Stakeholders in this instance include harvesters, but also other groups. Such rent-seeking behaviour is often the cause of difficulty or even failure in the implementation of rebuilding plans.

Objectives set as part of the plan must be realistic and attainable, which calls for the plan itself to be enforceable. Stakeholder involvement in the rebuilding plan can facilitate enforcement by their willingness to commit to the plan and the necessary actions.

Reforms in fisheries management will, in most cases, affect different stakeholders in different ways, raising a host of distributional issues. One consideration is how costs and benefits from the rebuilding plan are distributed over time and among stakeholders. The costs of rebuilding fisheries tend to be upfront and immediate, while the associated benefits may take time to appear and be shared by many. That means that those who initially bear the cost of the rebuilding plan may not be the ones who reap the future benefits. In order for fishers to be willing to make sacrifices, they must expect a reward for their efforts. Otherwise they will not have an incentive to take part in the rebuilding effort. Policy makers must consider this issue in the design of rebuilding plans. They have various options to mitigate such risks, e.g. by using transfers which depend on realised outcomes. In some cases, it will be necessary to use flank measures and compensation schemes, not only to compensate for lost revenues, but also to ensure the success of the reform itself (Box 5.2).

Identifying the stakeholders and how they will be affected by the rebuilding plan is necessary for success. Experience shows that when stakeholders are passive it weakens management plans. It is necessary to map out not only who the stakeholders are, but also what role each will play (if any). Who, how and when decisions are taken are issues that must be addressed. Although underlines the importance of incorporating stakeholders, where appropriate, in rebuilding plans it must be kept in mind that rebuilding a fishery is in many cases similar to a public investment project where final decisions and responsibilities lie with the public authorities.

Several case studies for this report underline the importance of stakeholder participation in the whole rebuilding process, from preparation to implementation. In many cases, stakeholders initiated rebuilding efforts, such as in Japan, Korea, Iceland and France. The Sailfin sandfish in Akita prefecture in Japan is an interesting case where stakeholders have an active role in the rebuilding effort. After decades of input controls and technical regulations, the state of the stock resulted in a fishing ban from September 1992 to September 1995. This was a *self-imposed ban* by local stakeholder groups. After the three-year ban, a TAC was implemented. The TAC was set by a local fishery management group at half the estimated spawning fish population. Although the biological situation at present is better than before, the current economic situation is not due to falling prices and the introduction of new species that have replaced sailfin sandfish in the market.

This case study was developed according to two objectives: a) to ensure biological recovery and, b) maximise profits for individual fishers. There was an active dialogue between the national and municipal governments, and local stakeholder groups. It is important to note that the limited entry into this fishery guaranteed that expected benefits were received by the same fishers who bore the costs of the ban. Stakeholder participation was high which explains the self-imposed regulation in Japan's coastal fisheries. There was considerable self-enforcement. A compensation scheme was used to distribute the burden of the rebuilding effort equally among the national and prefectural governments and the fishers.

Box 5.2. Mitigating distributional effects

The ability of management to mitigate the distributional effects of rebuilding relates to its ability to articulate a full range of policy choices, a time path for rebuilding, and clear rules for decision-taking. Stakeholder involvement in defining policy choices influences their acceptability, but this involvement rests in turn upon the tenure assurance that allows engagement in long-term planning. Communities are included in the group of stakeholders needing tenure assurance.

The time path for rebuilding is often contentious, especially with long-lived species that are caught in a mixed-stock fishery. In these cases, the distributional effects are compounded over time and over species, providing greater incentives for stakeholders to resist rebuilding. Possibilities for mitigation may rest on the existence of substitute fishing activities. Without substitution possibilities, a clearly defined time path for rebuilding combined with tenure security for fishery participants may provide a long-term possibility for mitigation.

A rule-based approach for rebuilding requires precautionary or limit reference points to be defined and a decision about non-discretionary action that will be taken if these limits are reached. Experience in stock rebuilding shows the importance of specifying the conditions under which rebuilding is obligatory and strict enforcement of these during the rebuilding process (Caddy and Agnew, 2004; FAO, 2005).

Source: Hanna (2009).

Decision on mechanisms to rebuild fisheries

Theory and experience show that properly designed rights-based management (RBM) systems may be effective in encouraging responsible behaviour through incentives and thereby support the rebuilding process. The experiences of the New Zealand hoki fishery and the Icelandic pelagic fisheries are examples of this. However, RBM systems, such as ITQs, have often been criticised for focusing on profits and economic rent rather than ecological objectives, such as biodiversity, or wider social objectives, such as employment, social justice or cultural heritage. While correctly designed RBM systems do maximise the social benefits that can be derived from the resource, it is true that some of the possible benefits do not have a market value and would therefore not be accounted for in the optimisation process working through market transactions.

There is often political resistance to the introduction of RBM systems. The general notion that fish resources are and should be common property is widespread, although it has been clear for a long time that most fisheries suffer when lacking well defined exclusive rights. Fisheries as common property are also often prescribed in legislation, which may make it difficult for policy makers to move towards RBM. It should be noted that resistance or acceptance of RBM hinges to some extent on the party or parties to whom the rights are ascribed. For example, community-based rights schemes are likely to garner less resistance from some stakeholders.

It is noteworthy that most RBM systems do not privatise the resource, but rather create exclusive user rights to fishers. In that way the resource itself remains common property while the right to use it is individualised.

Exclusive fishing rights exclude parties other than right-holders from taking part in the fishery. How fishing rights are or can be distributed is often a heated political issue and is one of the weakest facets of RBM plans in many countries. Most often such exclusive rights are granted on the basis of catch history (i.e. a grandfathering system), but this is seen by many opponents as essentially gifting a valuable publicly-owned resources to private interests. To overcome this objection, auctions have been used sometimes to distribute fishing rights, but this method is often resisted by fishers.

If fisheries managers and policy makers believe that RBM is not a feasible or desirable option, there are a range of other direct fisheries management measures available. Although it is questionable whether and when RBMs are "optimal" from a welfare point of view, they can be effective in rebuilding fish stocks and generating some rent. It should be noted that RBM systems rely on various direct (command and control) measures to function efficiently.

The use of licenses, such as in the Namibian hake fishery, and taxes, as in Mauritania, have been relatively successful in maintaining sustainability in certain fisheries. Such systems rely heavily on the ease of monitoring catches and experience shows that they are more easily introduced in industrialised fisheries than in artisanal fisheries.

Whether policy makers opt for indirect (incentive or rights-based) or direct (command and control) management policies in rebuilding relies on data availability and the cost of monitoring and surveillance. There are no direct links, however, between type of management (direct or indirect) and data intensity or cost of management (Larkin *et al.*, 2011). Data intensity and cost of management should be evaluated for each case by itself. Furthermore, all rebuilding policies contain a mix of direct and indirect management measures.

Finally, different management systems come at different costs and some are not easily implemented. Various management systems require extensive data on catches, effort and landings. If such data are unavailable or too costly to collect, fisheries managers may be forced to choose another type of system even though it may have a lower probability of success. Data collection can be an important part of a rebuilding plan when the gains from rebuilding are demonstrated and thus contribute to funding. Most governments have fiscal constraints which normally affect the decisions taken on fisheries management. This underscores once again the importance of policy choices regarding rebuilding fisheries plans.

References

Caddy, J.F. and D.J. Agnew (2004), "An Overview of Recent Global Experience with Recovery Plans for Depleted Marine Resources and Suggested Guidelines for Recovery Planning", *Review of Fish and Fisheries*, Vol. 14, pp. 43-112.

FAO (1995), *Code of conduct for responsible fisheries*. Food and Agriculture Organisation of the United Nations. Rome. FAO (2002), *CWP Handbook of Fishery Statistical Standards*. Section G: FISHING AREAS - GENERAL. CWP Data Collection. In: FAO Fisheries and Aquaculture Department [online]. Rome. Updated 10 January 2002. [Cited 12 December 2011]. *www.fao.org/fishery/cwp/handbook/G/en*

Hanna, S. (2009). "Managing the transition: distributional issues of fish stock rebuilding" in OECD Workshop Proceedings: Economics of Rebuilding Fisheries.. Paris..

Larkin, S., S. Alvarez, G. Sylvia, and M. Harte. (2011), "Practical Considerations in Using Bioeconomic Modelling for Rebuilding Fisheries", *OECD Food, Agriculture and Fisheries Working Papers*, No. 38, OECD Publishing. http://dx.doi.org/10.1787/5kgk9qclw7mv-enLee, Sang-Go (2009). "Rebuilding fishery stocks in Korea: a national comprehensive approach", in OECD Workshop Proceedings: Economics of Rebuilding Fisheries. Paris.

OECD (2011). *Fisheries Policy Reform. National Experiences*. Paris.

Sutinen, J.G. (2008). Major Challenges for Fishery Policy Reform: A Political Economy Perspective, *OECD Food, Agriculture and Fisheries Working Paper No. 8*, available online at www.oecd.org/fisheries.

ORGANISATION FOR ECONOMIC CO-OPERATION AND DEVELOPMENT

The OECD is a unique forum where governments work together to address the economic, social and environmental challenges of globalisation. The OECD is also at the forefront of efforts to understand and to help governments respond to new developments and concerns, such as corporate governance, the information economy and the challenges of an ageing population. The Organisation provides a setting where governments can compare policy experiences, seek answers to common problems, identify good practice and work to co-ordinate domestic and international policies.

The OECD member countries are: Australia, Austria, Belgium, Canada, Chile, the Czech Republic, Denmark, Estonia, Finland, France, Germany, Greece, Hungary, Iceland, Ireland, Israel, Italy, Japan, Korea, Luxembourg, Mexico, the Netherlands, New Zealand, Norway, Poland, Portugal, the Slovak Republic, Slovenia, Spain, Sweden, Switzerland, Turkey, the United Kingdom and the United States. The European Union takes part in the work of the OECD.

OECD Publishing disseminates widely the results of the Organisation's statistics gathering and research on economic, social and environmental issues, as well as the conventions, guidelines and standards agreed by its members.

OECD PUBLISHING, 2, rue André-Pascal, 75775 PARIS CEDEX 16
(53 2012 01 1 P) ISBN 978-92-64-17692-8 – No. 60097 2012-01